BestMasters

Springer awards „BestMasters" to the best master's theses which have been completed at renowned universities in Germany, Austria, and Switzerland.

The studies received highest marks and were recommended for publication by supervisors. They address current issues from various fields of research in natural sciences, psychology, technology, and economics.

The series addresses practitioners as well as scientists and, in particular, offers guidance for early stage researchers.

Thorsten Will

Predicting Transcription Factor Complexes

A Novel Approach to Data Integration in Systems Biology

 Springer Spektrum

Thorsten Will
Saarbrücken, Germany

BestMasters
ISBN 978-3-658-08268-0   ISBN 978-3-658-08269-7 (eBook)
DOI 10.1007/978-3-658-08269-7

Library of Congress Control Number: 2014956553

Springer Spektrum

Springer Spektrum is a brand of Springer Fachmedien Wiesbaden
Springer Fachmedien Wiesbaden is part of Springer Science+Business Media
(www.springer.com)

Geleitwort des Betreuers

Die genetische Information jeder biologischen Zelle ist bekanntlich in deren Erbsubstanz kodiert. Insofern gibt es beispielsweise keinerlei Unterschiede zwischen den etwa 200 verschiedenen Zelltypen eines Menschen (Nervenzellen, Muskelzellen, Knochen, Haut etc). Obwohl diese Zellen natürlich große morphologische Unterschiede aufweisen, enthalten alle Zellen im Prinzip dieselbe Information. Entscheidend für den jeweiligen Zustand einer menschlichen Zelle - d.h. deren Differenzierung in die einzelnen Gewebetypen - ist deshalb nicht, welche Informationen die Erbsubstanz prinzipiell enthält, sondern welche der etwa 22.000 Gene in ihr tatsächlich „abgelesen" werden und welche nicht. Man kann dies vereinfacht mit dem Lesen eines Buches vergleichen, bei dem auch jeweils eine Seite aufgeschlagen wird. Wer daher in der Zelle diesen Lese-Prozess reguliert, bestimmt quasi das Schicksal der Zelle. Die Entschlüsselung dieser Prinzipien zur Regulierung der Gentranskription ist somit ein sehr wichtiger Schlüssel zum Verständnis von Zellen.

Die wichtigste Rolle bei der Entscheidung, welche Gene abgelesen werden sollen, übernehmen Eiweißmoleküle, sogenannte Transkriptionsfaktoren, die gezielt an Bindungsstellen an der Erbsubstanz binden können. Wenn diese Transkriptionsfaktoren direkt „vor" dem Beginn einer Gensequenz an die Erbsubstanz binden, kann der Kopierapparat der Zelle, die RNA-Polymerase, gezielt an diese Region rekrutiert werden und beginnt, die sich anschließende Region der Erbsubstanz zu kopieren, d.h. abzulesen.

Man könnte sich nun vorstellen, dass zu jedem einzelnen Gen ein bestimmter Transkriptionsfaktor gehört, der den Ableseprozess dieses Gens reguliert. Jedoch müsste ja auch dieser Transkriptionsfaktor in der Zelle durch Ablesen eines anderen Gens produziert werden, so dass man dann zweimal so viele Gene bräuchte. Und so weiter. Man sieht leicht, dass solch eine Variante nicht funktionieren kann. Eine andere Möglichkeit wäre, dass ein einzelner Transkriptionsfaktor jeweils das Ablesen von vielen anderen Genen reguliert. Dann käme man mit einer wesentlich geringeren Anzahl an Transkriptionsfaktoren aus, könnte aber die Regulation der einzelnen Gene nicht mehr so feinkörnig steuern. Eine dritte Möglichkeit, die nun tatsächlich in Zellen realisiert wird, ist dass jeweils mehrere Transkriptionsfaktoren gemeinsam den Ableseprozess von einzelnen Genen kontrollieren. Damit reichen wenige hunderte an Transkriptionsfaktoren aus um eine enorme Anzahl an kombinatorischen Varianten zu erzeugen.

Der Autor entwickelte in seiner Masterarbeit im Fachgebiet Bioinformatik einen neuen Ansatz um Eiweißkomplexe zu identifizieren, die aus mehreren Transkriptionsfaktoren sowie aus weiteren Proteinen bestehen. Um die Praktikabilität der Methode zu testen, wurde als Modellorganismus die Bäckerhefe (S. cerevisiae) ausgewählt, da hierfür besonders gute experimentelle Daten zu paarweisen Proteininteraktionen vorliegen. Der Algorithmus verwendet die in der Informatik oft eingesetzte Baumstruktur zur Aufzählung aller möglichen Komplexe, die an die Erbsubstanz binden und maximal 10 Eiweißmoleküle enthalten. Ein neuartiger Beitrag bestand darin, die grundlegende Datenstruktur für die Interaktionen nicht auf gesamten Proteinen aufzusetzen, sondern auf deren Domänenbausteinen. So konnten zum einen weitere Interaktionsdaten zwischen Proteindomänen eingebunden werden. Zum anderen ergibt sich eine feinere strukturelle Auflösung der in Konkurrenz miteinander stehenden Kontakte. Mit dem neuen Ansatz konnten für Hefe mehr als 10 mal so viele unterschiedliche Proteinkomplexe generiert werden wie mit anderen derzeit verfügbaren Methoden. Der Autor zeigte zudem, dass die Ergebnisse eine bessere Abdeckung der bisher experimentell charakterisierten Komplexe liefern als alle anderen Methoden und dass die vorhergesagten Komplexe eine hohe biologische Plausibilität besitzen.

Diese vielversprechenden Ergebnisse lassen es denkbar erscheinen, ähnliche Methoden auch für komplexere Lebewesen wie die Maus oder sogar den Mensch einzusetzen. Dies wäre ein wichtiger Schritt dabei, die Mechanismen der Genregulation besser zu verstehen, da deren Fehlfunktionen natürlich auch zur Entstehung vieler Krankheiten beitragen. Die Bioinformatik übernimmt bei solchen Projekten meist die wichtige Aufgabe der Datenintegration und ermöglicht es, die Existenz bestimmter Szenarien oder Mechanismen zu postulieren, deren Korrektheit dann im Experiment gezielt getestet werden kann.

Prof. Dr. Volkhard Helms

Institutsprofil

Das Zentrum für Bioinformatik an der Universität des Saarlandes[1] ist eine interfakultäre Einrichtung zwischen der Informatik, Medizin und den Lebenswissenschaften Biologie, Chemie und Pharmazie. Ein wichtiger Schwerpunkt der Saarbrücker Bioinformatik-Forschung ist die medizinische Bioinformatik, d.h. die Anwendung von Bioinformatik-Methoden für die Bearbeitung von biomedizinischen Daten. Man interessiert sich heutzutage zum Beispiel dafür, wie sich die Erbsubstanz in Krebszellen von der Erbsubstanz in daneben liegenden gesunden Zellen unterscheidet. Ein anderes wichtiges Forschungsgebiet ist die Resistenzforschung, wie es Viren und Bakterien durch Veränderung ihres Erbgutes schaffen, die Wirkung von an sich hoch aktiven antiviralen oder antibakteriellen Wirkstoffen ins Leere laufen zu lassen.

Das Zentrum für Bioinformatik organisiert an der Universität des Saarlandes einen grundständigen Bachelorstudiengang Bioinformatik und einen darauf aufbauenden Masterstudiengang. Basierend auf den anerkannten Forschungserfolgen, aber auch befördert durch die Anbindung an die exzellenten Forschungsinstitute der Saarbrücker Informatik und durch die hohe gesellschaftliche Relevanz der bearbeiteten Forschungsthemen, genießt die Saarbrücker Bioinformatik weltweit eine hohe Reputation. Die Absolventen der beiden genannten Studiengänge sind in nationalen und internationalen Unternehmen sowie in Forschungsinstituten sehr stark nachgefragt.

---

[1]www.zbi.uni-saarland.de

Preface

Gene regulatory networks are fixed determinants of cellular control and the abundance of differentially expressed regulatory proteins, called transcription factors, their driving signal. In concert with specific epigenetic marks, transcription factors define the active subset of the network to govern distinct cellular states in time and space.

Eukaryotic gene expression is generally controlled through molecular logic circuits combining regulatory signals of several transcription factors. Recently, it has been shown that complexation of regulatory proteins is a prevailing and highly conserved mechanism of signal integration within critical regulatory pathways, like body part formation or differentiation. A knowledge of potential assembly candidates could provide the basic information that is needed to infer possible target genes as well as the exerted mechanism of influence. There already exists a plethora of approaches to predict protein complexes from protein-protein interaction data. However, those are generally designed to detect large self-contained functional complexes and lack the ability to reveal dynamic and highly modular combinatorial complex assemblies, a property of crucial importance for the signal integration exerted by transcription factor complexes.

The method proposed in this thesis combines protein-protein interaction networks and domain-domain interaction networks with the well-known cluster-quality metric cohesiveness. A novel growth algorithm is described that locally maximizes the metric on the holistic level of protein interactions while sophisticated connectivity constraints are preserved. Assuming that each domain can only support one interaction, the domain topology can be utilized to account for the exclusive and thus combinatorial nature of physical interactions between proteins. During the growth process, the complex candidate is thought to be backed by a spanning tree of simultaneously possible domain interactions which restrict further expansion possibilities. Consequently, every addition of a protein requires the choice of an applicable domain interaction which again influences later steps. Often many options have to be taken into account by branching of the algorithm, which naturally allows for the justified prediction of a manifold of transcription factor complexes from a common start.

The proposed approach outperformed popular complex prediction methods by far for the prediction of transcription factor complexes in yeast. The evaluation was based on established benchmarks assessing accordance with several reference complex datasets as well as measures of biological relevance. Additionally, many of the predictions of the proposed method could be associated with target genes and a potential regulatory effect. Furthermore, predicted candidates could be mapped to distinct functions during a defined cellular state and condition by analyzing the expression coherence among their regulated genes for cell cycle expression data. Many findings were backed up by literature evidence.

The results encourage an application to higher eukaryotes where the combinatorial interplay between transcription factors is more pronounced. The knowledge of putative transcription factor complexes - DNA-binding members and recruited potentially regulatory active proteins - offers novel capabilities in the automatized modeling of gene regulatory networks which may assist to surpass nowadays models.

A condensed summary of the novel concept, the main method and the results for yeast was previously published in [1] prior to the production of this book.

Contents

List of Tables

List of Figures

List of Algorithms

List of Abbreviations

bp   base pair(s)

CRM   cis-regulatory module

DDI(N)   domain-domain interaction (network)

EC(S)   expression coherence (score)

GO   Gene Ontology (annotation)

GTF   general transcription factor

HMM   hidden markov model

MMR   maximum matching ratio

ORF   open reading frame

PDB   Protein Data Bank

Pol II   RNA-Polymerase II

PPI(N)   protein-protein interaction (network)

SPIN   simultaneous protein(-protein) interaction network

TBP   TATA-binding protein

TF   transcription factor

TSS   transcriptional start site

YPA   Yeast Promotor Atlas

Introduction

1.1 Motivation

Although the information stored in the DNA sequence of every cell is the same, different cells in multicellular organisms differ dramatically. Especially higher eukaryotes exhibit various cell types and tissues; even two adjacent cells are never completely identical. The phenotype of a cell is determined by the composition of its expressed proteins which changes during its life cycle, among cell types and in response to physiological and environmental conditions. Many essential processes are common to all cells, like the basic pathways of metabolism, maintenance mechanisms or the preservation of structural integrity. Proteins that are involved in such functions are ubiquitously expressed across different cell types and tissues. Some proteins in turn are only abundant in very specialized cells and never found somewhere else, such as hemoglobin in red blood cells, contractile proteins in muscle cells or surface receptors of immune cells [2–4].

Surprisingly, the advent of whole-genome sequencing accompanied by comparative genome analysis showed that morphological complexity does not correlate with the sheer number of genes. On average, vertebrates have only twice as many genes as invertebrates, while many of those are even duplicates [5]. The fruitfly *Drosophila*, for example, has fewer than 14, 000 genes, but many more cell types and tissues as seen in the nematode worm *Caenorhabditis elegans* that has almost 20, 000 genes [6]. Meanwhile, the likely number of gene expression patterns, the "regulatory complexity", is assumed to be the driving force behind the physiological and behavioural diversity within cells across all plants and animals [3, 6].

While the basic pathway from DNA to protein ("DNA is transcribed into mRNA, mRNA is translated into proteins") is a rather concise one [7], there are many possibilities to regulate the abundance of the active form of a protein. The control starts at the level of transcription, involves splicing, processing, localization as well as degradation of RNA transcripts, the regulation of translation and the surveillance of protein activity, by posttranslational modifications, localization and degradation [2].

Although all of these mechanisms are important for the regulation of active genes in general, the early steps in a linear control cascade are usually the paramount ones: repression or absence of intermediates at a certain regulatory step will render the later ones redundant. This is especially important for cell type-specific activity of proteins and makes transcriptional regulation the most important control point of differentially expressed genes. Moreover, from an economical point of view it makes sense to steer protein composition primarily at the level of transcription because only that will ensure that no unnecessary intermediates are synthesized [2, 3, 8]. For this reason, we will only consider the regulation of transcription for this thesis.

While less than 2% of the human genome encodes proteins, up to a third of the genome is believed to serve regulatory purposes [6]. In humans, up to 10% of the protein-coding genes are associated with transcriptional regulation. These proteins can be categorized into three major classes: transcription factors (TFs), DNA-binding regulatory drivers which target regulatory regions of specific genes, functional proteins of a basal transcription machinery and proteins that are able to influence the packaging of the DNA [6].

DNA is hierarchically organized as chromatin, a complex of bare DNA and proteins. On the lowest level, small segments of DNA (less than 150 base pairs (bp)) are wrapped twice around a unit of eight histone proteins to form a nucleosome, which again can be further compacted to solenoids. Such condensed regions are less accessible for DNA-binding proteins which constitutes another major control layer of cellular state [8]. The packaging, on one hand, prevents leaky transcription in a directed way, allowing eukaryotes to exhibit an up to 10^6 fold variation in expression (bacteria: only up to 1000 fold), but can also be passed on by a cell to its descendants, facilitating a medium/long-term memory for expression control [2]. The local condensation state of the chromatin is controlled by epigenetic mechanisms like posttranslational modifications of the histones by special reader and writer proteins as well as methylation of cytosines within the DNA sequence, both mechanisms which are in a steady interplay among each other and can be influenced by TFs [8–11].

The expression of every gene is controlled by a set of TFs that bind to specific regulatory regions, those TFs in turn are proteins coded by genes that are again controlled by an entire set of other regulatory proteins, leading to a whole network of directed dependencies. Such networks are called gene regulatory networks [3].

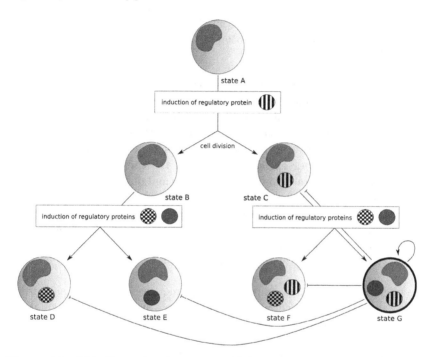

Figure 1.1: This illustration shows a simplified scheme how combinatorial control allows to generate many different cellular states with only a few regulatory proteins. Few differentially expressed TFs can facilitate the specification of multicellular organisms. The initial state (state A), the fertilized oocyte, consists of maternal regulators which are required to start the process. From then on, anisotropies in cell divisions, discrete events of stochastic activation or external signals lead to differential expression of key regulator genes [2, 3, 12]. Stable cellular states, characterized by their individual composition of key regulators (for example state G, marked thick), are stabilized by self-perpetuating feedback-loops and foster active exclusion of alternative regulatory states [2, 3].

Figure 1.1 explains how the combinatorial control of few genes can in principle give rise to exponentially many regulatory states in such a network. If states in the tree are equated to cell types in the developmental regulatory networks of higher eukaryotes, every stochastic decision event may be considered as a lineage commitment [12]. With every further specification step, specific epigenetic lockdown mechanisms are deployed. In contrast to the transient physiological signals of protein expression, those allow the cell line to establish a long-term memory of the state that is even inherited to the mitotic descendants. The thus imprinted epigenetic pattern indicates the remaining differentiation potential. The terminal nodes in such trees then represent terminally differentiated cells which express the different phenotype-specific genes [3, 9, 13].

Gene regulatory switches that toggle such important decisions of cellular fate exploit the principle of combinatorial control with elaborate molecular logic circuits. They generally require the abundance of several distinct key regulators altogether with a certain chromatin state of the affected genomic region [3, 9, 13]. This strictness is necessary, since even small deficiencies within gene products or epigenetic marks in the complex interdependencies of regulatory interactions can entail abnormal development and severe diseases, like cancer [3, 9, 14, 15]. A 'correct' combination of factors, on the other hand, may induce striking reactions from the network, like the reprogramming of actually differentiated somatic cells by a reset of the differentiation potential. Such techniques may at some point create a new era of personalized regenerative medicine [13].

The understanding of the complex interplay requires a paradigm shift from single key players to the collaboration of a whole system, where regulatory sequences and proteins provide an initial causal link.
A detailed view on eukaryotic transcriptional regulation will expose the major principles of interaction and signal integration between all parties involved and finally suggest a novel computational strategy to obtain a good idea of potential key players in combinatorial control circuits.

1.2   Eukaryotic transcriptional regulation

Every eukaryotic gene with a phenotypic impact contains a promotor, a regulatory sequence region located upstream (or 5') of the transcriptional start site (TSS) which, together with the currently active regulatory proteins encoded elsewhere and the state of the chromatin, determines to which extend expression of the gene occurs. Flanked by random DNA, the gene would be inert [2, 4, 8].

A promotor harbours diverse binding sites for special DNA-binding proteins, the transcription factors (TFs) that are used to integrate information about the current status of the cell, the physical or environmental condition but also tissue and cell type, to regulate a transcriptional response for each individual gene accordingly. Polycistronic transcripts, like operons in prokaryotes, are uncommon in eukaryotes [3, 4]. The typical size of a promotor sequence spans magnitudes from hundreds of base pairs (bp) in yeast up to hundreds of thousands bases for a few genes in mammals [2, 4, 16].

Figure 1.2 shows the functional segmentation and binding site distribution of a typical eukaryotic gene. The TSS separates the gene into the part of the sequence that is transcribed and the promotor. Moreover, the promotor can be further divided into a proximal and a distal part.
The TSS is determined by the basal or core promotor, a sequence segment with characteristic base pair motifs spatially near (< 100 bp distance) to the TSS that are important for the initiation of transcription. Adjacent to the core promotor in upstream direction are the proximal promotor elements, an accumulation of few transcription factor binding sites that are used to influence the process of transcription initiation. The remaining part is called the distal promotor [4, 8, 17, 18]. The functional role of and mechanisms behind all these individual components will be explained in the upcoming sections.

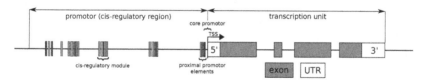

Figure 1.2: Architecture of a common eukaryotic gene showing the relative position of the transcriptional unit, the proximal promotor adjacent to the TSS (that is defined by the core promotor) and transcription factor binding sites (colored vertical bars). The distal promotor comprises the region 5' of the proximal promotor that is more distant to the TSS. Binding sites are sparsely and unevenly distributed, often overlap and accumulate in clustered dense areas, termed *cis*-regulatory modules. For completeness, the transcribed sequence is further divided into untranslated regions (UTRs) and exons and introns, the building blocks used to form proteins in splicing events [2, 4, 8]. The continuous text will elaborate some more details.

Transcription initiation

The majority of genes has a sequence element called the TATA-box (TATA) on the core promotor which serves as the binding site for the TATA-binding protein (TBP). The TATA-box is located about 25-30 bp upstream relative to the TSS and serves the alignment of a protein machinery that designates the TSS. The actual TSS has no distinctive sequence motif [2, 4]. Some genes lack the TATA-box and have different initiator elements that facilitate TBP association in a sequence-independent manner and there are also null basal promotors that have neither one. The functional differences between TATA/TATA-less promotors are not well understood, but TATA is often absent in housekeeping genes that are universally expressed [2, 6, 8].

Binding of TBP leads to the recruitment of other proteins, the general transcription factors (GTF), and finally RNA-Polymerase II (Pol II), the polymerase responsible for the catalysis of transcription of DNA to synthesize precursor mRNA. Other polymerases are responsible for ribosomal RNA and various small RNAs [2]. The pathway of assembly is not strict for this so-called basal transcription (or preinitiation) complex, some GTFs can even be preassembled with Pol II forming a holoenzyme [8].

The GTFs are sometimes referred to by their functional subunits, abbreviated as TFIIA, TFIIB and so on. TBP is a part of TFIID, which is the largest component with about 12 proteins [2, 6]. Overall plenty of proteins are involved in a basal transcription complex and even more can be a component, like the Mediator cofactor complex (to be explained soon) and various chromatin modifying or remodelling complexes (RSC and SWI/SNF in yeast, many more in higher eukaryotes). Together, they ensure the proper positioning of polymerase at the TSS, aid in pulling apart the two DNA strands and pave the path for transcription [6, 8, 19].

After successful assembly and positioning, Pol II is released into its elongation mode by the GTFs and the expression of the gene begins. As Pol II moves downstream to read the sequence, TFIIF remains bound to the polymerase while TFIIA and TFIID stay associated with the core promotor and are thus able to recruit another molecule of polymerase for a repeated round of transcription [8]. Figure 1.3 visualizes and summarizes this cyclic procedure.

However, the level of expression generated by the sole basal machinery is functionally not significant. Since all participating proteins are ubiquitously expressed within an organism this machinery is also not suitable for

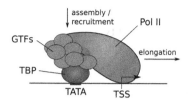

Figure 1.3: Transcription works by an initial assembly of a multiprotein complex and repeated iterations of Pol II recruitment and release into elongation. An acceleration of these two steps leads to a higher rate of transcription [8].

regulation. For a strong signal as well as specific control the recruitment needs to be enhanced further by an important class of regulatory proteins, the transcription factors (TFs) [4].

1.2.1 Transcription factors

Transcription factors (TFs) are DNA-binding regulatory proteins with a strong effect on the transcriptional activity of influenced genes. They mainly use the major groove of the DNA to establish a sequence-specific contact since the exposed hydrogen bonds therein provide accurate information about the corresponding nitrogenous base. Specific TF binding sites typically encompass only 6-12 bp, which is a fairly low number and suggests more complex additional rules that will be of major interest later [2, 4, 20]. Common TFs consist of a few very well conserved, yet separable, functional modules [4, 21]:

DNA-binding domains
facilitate the binding capability, may be continuous or dispersed within the primary amino acid sequence. Some TFs contain two distinct binding regions.

activation domains
enable protein interaction which will be used to carry out a certain regulatory function or to recruit other regulatory proteins. Such domains in a wider sense enable the active influence on the rate of transcription.

ligand-binding domains
enable to react to external stimuli, which can lead to structural changes that can influence binding possibilities and activity.

trafficking domains
steer the import from cytoplasm to nucleus to regulate the active form.

TF gene families are generally classified by their DNA-binding domains. Well defined domains are, for example, the helix-turn-helix motif found in homeobox transcription factors or various zinc finger motifs. Overall, only 12 to 15 distinct DNA-binding domain families are known for eukaryotic transcription factors and those may represent a nearly complete list for intensively studied organisms [4, 22].

While DNA-binding domains are well characterized by their structural folds and can also be detected from sequence, activation domains are much harder to grasp and few of them are structurally determined [21, 23].

Apart from ubiquitous base mutations, the evolutionary diversity within TF families can often be explained by shuffling or loss of specific domains. Thus the interchange (or redistribution) of a few functional modules can yield a great variety of different TFs. For example, a paralog (a gene that emerged from a duplication event but exhibits altered function) may keep the same DNA-binding domain, but forfeits the activation domain. The paralog will then compete for the same binding sites with other paralogs that also have the DNA-binding domain, but indirectly act as a repressor by replacing those paralogs that still contain the domain responsible for the activating function [4, 21, 24].

The remaining section will be dedicated to the modes of action exerted by such activation domains in TFs.

TFs can influence transcriptional initiation and elongation

TFs with binding sites near to the TSS in the proximal promotor are often able to directly interact with GTFs through protein-protein interactions enabled by their activation domain. Acidic activation domains, for example, preferentially interact with TFIID to stimulate its binding to the promotor [8]. Such activators can increase the rate of complex assembly by independently enhancing several steps in the construction of the basal transcription complex as well as by fostering the recruitment of polymerases and their elongation. If TFs that promote several of those steps are present, these effects add up and lead to a strong synergistic upregulation of a gene's transcription [2, 4, 8].

For eukaryotic TFs it is often the case that protein interactions communicated by their activation domain(s) do not solely aim to contact GTFs directly but are used to bind other proteins, in general called cofactors, to build relevant functional complexes. Cofactors in this context are proteins that have no DNA-binding ability themselves but can mediate between TFs and regulatory important proteins like the GTFs or proteins that in another way contribute to expression control. If they directly govern the decisive regulatory function and are not only used as adapters, they are more specifically referred to as coregulators. Coregulators which lead to an increase in the rate of transcription are called coactivators, factors that cause a decrease are consequently termed corepressors [4, 20, 25, 26]. The need to have several proteins around simultaneously to accomplish a certain regulatory effect and the functional outcome's dependency on the final protein complex assembly of all individual components adds an additional layer of regulatory complexity [2, 4]. This context-dependency will be a central issue later on in the introduction.

Figure 1.4 shows a simplified illustration of the basic principles introduced so far.

Not all regulatory proteins with binding sites in the proximal promotor function as classical activators as described previously. Some can serve as tethering elements (or looping factors) that bend the DNA and are thus able to recruit distal parts of the promotor into spatial proximity to the core promotor and the basal machinery. Such looping factors are good examples for cooperative DNA-binding proteins that are not restricted by obvious distance constraints between binding sites [4, 6, 20]. This specific

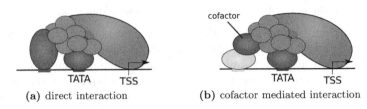

(a) direct interaction   (b) cofactor mediated interaction

Figure 1.4: Many TFs can directly influence the assembly and activity of the basal machinery to increase the rate of transcription by shortening the assembly/elongation cycles introduced priorly. They achieve this by **(a)** direct interactions with GTFs or by **(b)** indirect interactions that are mediated by cofactors, auxiliary proteins that also need to be present to obtain the transcriptional response.

deformation of the three dimensional structure enables TFs that bind far away in the sequence to also influence the preinitiation complex. Very important for such distal influences is the Mediator cofactor complex (over 20 proteins) which serves as a linker between the basal machinery and distant TFs by offering a huge binding area and relaying the incoming signals. Mediator is essential to control the transcriptional response to many regulatory inputs [2, 4, 27].

Figure 1.5 graphically visualizes the refinements just introduced to our model.

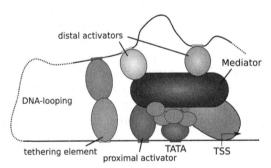

Figure 1.5: Looping of the DNA, bent and stabilized by assistant proteins, enables the usage of distal promotor regions for the regulation of transcription by interaction with the preinitiation complex. To allow the integration of various signals, the Mediator cofactor complex functions as a hub to provide access for the additional regulatory elements.

Eukaryotic cells also make use of repressive signals to modulate gene expression. The effect of downregulation can be induced indirectly by masking binding sites of activating TFs or masking their activation domains, but also in an active way. Active repression in this context is facilitated by TFs that inhibit recruitment steps in the assembly of the basal machinery, for example by sterical hindrance, or by blocking the release of polymerase into transcriptional elongation [2, 8]. In the total picture of regulatory networks, a TF A may also lead to an increased expression of a TF B which in turn acts as a repressor of gene C making A an indirect downregulator of C [4].

TFs can influence the chromatin structure

As already mentioned earlier, the interplay of TFs and chromatin state is a crucial regulatory control mechanism which eventually governs the course of differentiation in eukaryotes.
Besides their influence on the basal apparatus - by direct promotion of the assembly and elongation or indirectly by bending the DNA - another major mode of action exerted by TFs is the spatial control of the chromatin packaging. A TF bound to the DNA can locally influence the chromatin structure of a promotor or even specific areas within a promotor by recruitment of respective coregulatory proteins and multiprotein complexes as shown in Figure 1.6 [4, 26, 28].

Among those coregulators are ATP-dependent chromatin remodeling complexes such as SWI/SNF or RSC, which are able to alter the local DNA accessibility by the movement, ejection or restructuring of nucleosomes, and proteins that modify the tails of histones covalently [4, 9].
The most important covalent posttranslational alterations of histones are the acetylation of specific lysine residues in their amino acid chain, catalyzed by histone acetyltransferases (HATs) and reversed by histone deacetylases (HDACs), and the methylation of particular lysine and arginine positions by histone methyltransferases (HMTs) and demethylation by histone demethylases (HDMs). The added acetyl group is thought to neutralize the positive charge of the lysine to weaken the interaction with the negative charge of the DNA and thus promotes the opening of the chromatin and accessibility of the sequence region. The transcriptional influence of histone methylation depends on the position of the modification and additionally on the number of methyl groups, since lysine can be mono-, di-

and trimethylated and arginine up to dimethylated. Other modifications involve phosphorylation, ubiquitination and others [2, 9].

The entirety of such chromatin marks is often referred to as the "histone code". The information indicated by those marks is read by functional proteins that affect downstream processes in the transcriptional regulation of the genes, but also DNA replication and repair, since local 'DNA unwrapping' is also necessary for every mitosis [10]. Well-studied readers of the histone code with significant effect on the transcription of the affected genes are, for example, Polycomb-group proteins, Trithorax-group proteins or HP1 (Heterochromatin Protein 1) [9].
The transition from proliferation into differentiation of muscle cells, driven by the myogenic TF MyoD, is a noteworthy example that does not require the interplay of several TFs. Several relevant genes in proliferating muscle cells are held in a silent state by MyoD's recruitment of Suv39h1, a H3K9-methyltransferase. Abrogation of the specific HMT toggles the transcriptional activation of the required muscle genes [29].

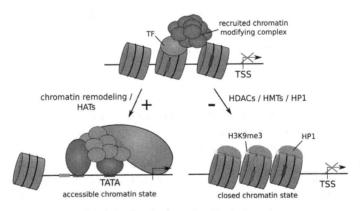

Figure 1.6: TFs are a driving force and targeting mechanism of chromatin state control by recruitment of corresponding proteins and protein complexes to precisely defined regions. Such an influence can effect a partially unwrapped state with increased protein accessibility (left) which could, for example, be caused by chromatin remodeling and proteins with histone acetyltransferase (HATs) activity. In contrast, a recruitment of histone deacetylases (HDACs) and special histone methyltransferases (HMTs) leads to a stable heterochromatin formation (right) strengthened by binding of HP1 proteins, which are able to read posttranslational modifications of histone tails and bind to trimethylated K9 positions of the histone H3 to enforce a locked down silent state [9].

Similar mechanisms have been shown for the interplay between TFs and DNA methylation. While methylated cytosines only impede the binding of some TFs, reader-proteins of DNA methylation marks - like MBP and MeCP2 - recruit heterochromatin-associated chromatin modifying proteins as mentioned before to cause a long-term compacted chromatin structure [9, 11, 30]. The exact targeting mechanisms of de-novo DNA methylation are still under steady investigation and could work by guided recognition of specific histone marks such as H3K9me2/3 and K3K27me3 [11]. Both de-novo DNA methyltransferases DNMT3a/b have been shown to interact with several histone reader-proteins suggesting a valid connection to histone modifications [30–32]. Such an interaction network of regulatory players supports the perception of a possible feedback loop mechanism that stabilizes silenced states [31].

1.2.2 Cis-regulatory modules

In prokaryotes, single TFs are able to regulate expression. Yet, this type of regulation is insufficient for eukaryotic transcription [2]. Here, the input signals of many regulatory 'drivers' - not necessarily restricted to the presence of specific promotor-binding TFs but inevitably also cofactors, epigenetic modifications or ligands - are integrated to regulate expression patterns of nearby genes context-dependent in time and space [20]. Especially genes encoding regulatory proteins often have complex expression profiles. "Housekeeping" genes, on the contrary, often only allow a shutdown due to extreme conditions like a heat shock [4].

This complex signal processing task is controlled by cis-regulatory modules (CRMs): functional DNA elements, typically 100-1000 bp in length, with clustered dense areas of binding sites (see Figure 1.2) in which each cluster contributes to the overall signal separately [4]. Their output is a combinatorial product of basic operations on the input such as conditional-, AND- and OR-logic implemented by an interplay of many physical interactions [3].

The phrase cis-regulatory (latin:"this side of") denotes their typical location on the same strand of DNA and relative vicinity to their regulated gene while trans-regulatory (latin:"far side of"), a term used rather rarely, refers to the influences by all regulatory elements on a different strand or farther away [3].

CRMs have often been called enhancers synonymously, but strictly speaking this includes only the modules that affect the expression of the controlled gene in a positive way. Other functional classes are silencers, which repress transcription, insulators, that limit the influence of other modules (likely by blocking the spread of chromatin alterations and looping), or locus control regions, modules in vertebrate genomes which can influence a set of genes by spatial approach [8, 18]. The net output of an individual module may be approximately boolean, whether a gene is turned on or off, or scalar/continuous, determining the rate of transcription. Often, separate modules are responsible for the initiation of transcription and its maintenance and actual activity [3].

The current view on their mechanics distinguishes between three putative prototype models: the enhanceosome, the billboard and the TF collective [20, 26]. Their commonalities and differences will be briefly addressed in this section, followed by a model-independent discussion of molecular mechanisms of action and consequences of cooperativity among TFs, a concept explaining non-linear contributions of regulatory inputs (see Section 1.2.2.1). Figure 1.7 will serve as a general overview of the models.

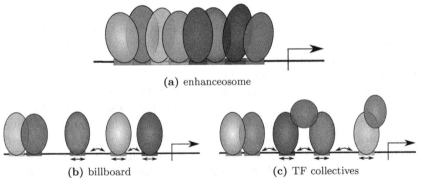

(a) enhanceosome

(b) billboard   (c) TF collectives

Figure 1.7: Current models of cis-regulatory signal integration range from the very strict and switch-like enhanceosome model, which assumes stringent constraints on pairwise distances and orientations for every TF, to the more flexible billboard and TF collective models, whose requirements regarding distances, orientations and even the composition of proteins are more variable and in some cases allow for a quantitative integration of the regulatory inputs [20, 26]. Double-headed arrows between binding site modules highlight indirect cooperativity, the arrows below binding sites depict unconstrained motif orientation.

Enhanceosome

The enhanceosome is a rare regulatory mechanism with very strict binding site constraints and may be limited to regulatory events that must be
controlled very precisely [18, 26]. The most-mentioned and best-studied
example is the transcriptional control of interferon-β, an immune response
protein [26, 33–35]. Many genes that are known to have an enhanceosome-
like control structure appear to be associated with pathways of wound
healing and pathogen defense [18].

For interferon expression, eight distinct TFs bind very tightly directly next
to each other in specific orientation and form direct pairwise protein interactions. They bind so tightly that their binding sites even slightly overlap
[35]. The typical distance between the binding sites of two tightly interacting TFs is shorter than 10 bp and also constrained by the helical structure
of DNA [36]. The high degree of cooperativity between the TFs leads to a
fast and sharp switch-like unitary signal to establish an adequate physiological response to viral infections [20].

This strictly-ordered arrangement of proteins leads to a very specific surface scaffold which is unattainable in solution at physiological concentra-
tions and is used by coactivator proteins like CBP/p300 and the GTFs to
form a functional complex that is very stable and even needs active dis-
assembly for removal and deregulation [34, 37, 38]. Usually a nucleosome
covers the TSS of the interferon-β gene and thus blocks the transcription
until the finalized complex achieves a removal of this blockade, catalyzed
by the recruited coregulators [39].

Complexes which interact in this strict fashion to repress transcription are
conveniently termed repressosomes [18, 40].

Billboard and TF collectives

Usually genes are organized in a more flexible fashion, better matching the
billboard and TF collectives models.

Observations made during the investigation of transcriptional regulation
of the *Drosophila* even-skipped (*eve*) gene, responsible for early body pattern formation, suggested that definite regulatory regions might be able
to represent both active and repressed states, depending on the cellular
context (external signals, cell type, TF abundance and more), and could,
taken together as the sum of individual contributions, modulate a continuous transcriptional signal [41]. Figure 1.8 illustrates the principle of
cis-regulatory integration that enables gradual transcriptional response in

the billboard model.

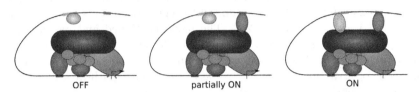

OFF   partially ON   ON

Figure 1.8: Contrary to the binary output of the enhanceosome's signal processing, the billboard model allows for a graded transcriptional response defined by the sum of discrete signal portions provided by individual and, perhaps, even contrasting submodules of TF binding sites which do not necessarily show any cooperativity among each other [20, 41].

Besides Figure 1.8, Figure 1.7b already summarized the essential conceptual changes in the architecture between the enhanceosome and the billboard model as proposed by [41]. While an enhanceosome demands not only exact distances but also a fixed motif orientation between every pair of binding sites to consistently allow for direct interactions between all TFs and only the presence of all involved TFs together is sufficient to produce a signal, the billboard model relaxes all those obligations just stated. In this model, an ensemble of regulatory regions interacts independently and/or redundantly to contribute a portion of an overall signal by exerting an activatory or repressory mode of action as described before. Subsets of factors may bind as cooperative units and not all factors are needed for a transcriptional signal. The motif grammar (relative positioning of binding sites within the promotor) across regulated genes is flexible, but the motif composition (appearance of each binding site within each promotor) fixed [38, 41].

The TF collectives model is the most recent one aiming to illuminate the convergence of decisive information within the promotors of tissue-specific genes. A genome-wide location analysis of five TFs that are essential for cardiac development in *Drosophila* suggested a new mode of action. Surprisingly, all five TFs are found at and regulate a large set of relevant enhancers even though each enhancer only harbours a diverse subset of motifs required for DNA-binding of all factors. Since the removal of only one TF already led to failing activation of all those enhancers in vitro, a model of consequent cooperativity and co-recruitment in vivo was suggested. Protein-protein interactions are thought to facilitate the collective occurrence of all TFs in the absence of any apparent motif grammar or shared motif composition. This provides an additional scaffold besides the regulatory sequence, an aspect further expanding the current understanding of cis-regulatory integration beyond the linear sequence-based code [20, 42].

Regardless of the exact control affecting a gene, the basic building blocks with relevance to the regulatory output are model-independent and worth a more precise look.
As mentioned earlier, even single TFs bound to the DNA (as in Figure 1.7c on the right) can change their regulatory function significantly if coupled to cofactors and, according to the TF collectives model, other TFs can, in principle, serve as cofactors and are not necessarily bound to the DNA in this function [8, 42]. Also, the exerted regulatory recruitment (the effect) and the binding specificity (the targets) of a TF can be modulated by conformational changes induced by plenty of factors like interaction partners, external stimuli by ligands or slight alterations in the DNA-sequence [6, 43, 44].
The cooperative interplay of several TFs can lead to particular decisive regulatory signals and is thus treated in a separate section.

1.2.2.1 Cooperativity between transcription factors

Where many TFs are working together tightly to enhance the rate of transcription, their combined effect is often not only the sum of individual contributions, but their synergistic product [2, 45]. Generally speaking, if a reactant lowers the free-energy barrier of a reaction accordingly to obtain a 10 fold speedup in the reaction rate and a second one does likewise, the combined speedup will be a factor of 100 [46]. In the context of transcription, where the rate is additionally limited by saturation, cooperativity between TFs leads to sigmoidal/switch-like output signals whose transition steepness increases with the degree of cooperativity [45, 47]. Regulatory proteins that collaborate within CRMs to produce such non-linear output signals and additionally provide logical operations are termed 'cooperative' [26, 45, 48].
Based on the pairwise distances between their binding sites, with the exception of connectivity via DNA looping, two cooperative TFs can be grouped roughly into three classes on the basis of the underlying mechanism. Direct interactions exhibit distances in the range of 0-10 bp and can show

additional bias regarding their motif orientations (in special cases overlaps are possible, as in the enhanceosome [35]), cofactor-mediated interactions typically are 10-50 bp away from each other [36, 44]. Non-bound cooperativity is also possible within binding site distances up to about 147 bp, corresponding to a stretch of DNA that is wound twice around a nucleosome. The synergy is then enabled by competitive interaction of the TFs with the same nucleosome [49].

A propensity for physical interactions between cooperative pairs is also reflected in protein-protein interaction data: cooperative TFs are closer together and more clustered within the protein interaction network as expected by mere chance with a significantly higher fraction of pairwise distances of one (direct) and two (mediated) within the graph [48, 50]. Also, they neither seem to share similar regulatory inputs nor regulate each other, although they influence similar groups of target genes, supporting their role as important regulatory drivers [48].

TFs very often bind as dimers or oligomers, like in the enhanceosome. If the type of interaction between them is a direct physical one, mediated by protein-protein interaction, this is called direct or classical cooperativity [26].

Building pairs (as in Figure 1.7c on the left), triplets or even higher-order tuples of TFs has many interesting consequences. When located in such complexes, even weak binders that are not able to accomplish this on their own can have an affinity that is strong enough to occupy a binding site stably [2].

Motifs of directly binding TFs in decondensed DNA are typically very near to each other (<10 bp), rely on precise spacing due to their size and the orientation matters [36]. Exceptions are interactions that take place through DNA looping and a few that achieve physical closeness by location on the same side of a nucleosome (around 40 bp) [4]. Since an oligomer requires two or more particular binding sites with correct distances (and in some cases orientation) between them, such a complex is way more specific than the binding of single TFs. Many recent publications show that cooperative binding is evolutionary much better conserved [36, 51–53] and many examples of polypeptide chains are known that have even been joined during evolution [2].

Another interesting and important point, however, is the intrinsic AND-logic such oligomers describe: for their specific sigmoidal signal all partners are needed, a conditional feature which automatized approaches for

gene regulatory network interference - with very few exceptions that re-
fine this issue unsupervised during a postprocessing step - are generally
lacking [3, 54]. Handtailored rule-based systems like boolean networks or
stochastic simulations can account for this [54]. The prediction of putative
TF complexes and possible roles and dependencies could be worthwhile to
automatically build a 'parts list' for such approaches.
As a consequence of the AND-logic, a single TF can still influence a whole
battery of genes by final completion or deconstruction of not necessarily
homogeneous regulatory complexes [2], comparable to the analogy of dial-
ing a final number of a combination lock. Figure 1.9 illustrates this basic
principle on an example.

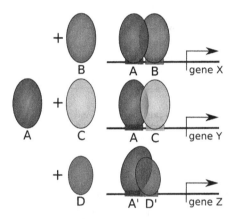

Figure 1.9: The addition of a sole TF A can toggle the expression of many genes. As
this example shows this can be facilitated by interactions with different partners. Additionally, some heterodimers (here A with D) can reveal latent specificity: an increased
binding specificity towards a, compared to the combination of their individual binding
motifs (A/D), slightly changed composite motif (A'/D'). [26].

The formation of TF complexes can also happen in an indirect manner
mediated by cofactors (as in Figure 1.7c in the middle). Such assemblies
are more flexible regarding their binding site limitations, pairwise distances
between binding sites are here in the range of 10-50 bp [36, 44]. As mentioned before, besides their functionality as an adapter, cofactors can serve
as necessary hubs for other regulatory proteins and can widely influence
the resulting regulatory mechanism [26, 44].
This rise in flexibility of combinatorics between binding partners by the

loosening of constraints together with the density of binding sites within cis-regulatory modules (see Figure 1.2) also allows to implement a certain OR-logic in promotors, since several combinations can lead to a certain equivalent signal [3].

While many TFs act primarily as activators or repressors, the definite effect on the rate of transcription depends on the final assembly of all its individual components [2, 4]. Figure 1.10 illustrates this and the OR-logic using a fixed set of regulatory proteins. As Lelli et al. put it in a recent review: the combinatorial possibilities are daunting [26].

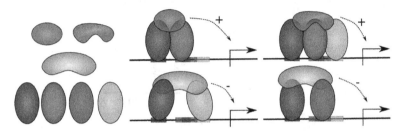

Figure 1.10: This example shows possible functional assemblies on the same regulatory sequence for a small set of putative TFs and cofactors. Some TF can participate in different complexes, even of opposing regulatory effect, and different assemblies in the same region can lead to the same net output signal.

Cooperativity is not inherently dependent on protein-protein interactions, though. While the basic building blocks of cooperativity so far are described to be working on "naked" DNA, cooperativity by "collaborative competition" occurs only on a chromatin template. Cooperative TFs that are not physically connected to each other can compete with a nucleosome for binding patches to their target sequences. Thermodynamic fluctuation of the affinity between nucleosome and DNA can allow one of the TFs to bind its target sequence with a certain probability and lower the interaction of the packaging which will then facilitate binding processes to neighboring patches of DNA within a region of 147 bp that has previously been wrapped around the nucleosome (see Figure 1.11) [26, 49].

Special TFs called "pioneer factors", like the family of FOX (forkhead box) proteins, are able to bind chromatin in the condensed state and are thus able to overwrite an epigenetic state and to provide access of other TFs to the previously sealed off DNA by collaborative competition [20, 26]. This

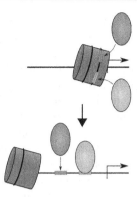

Figure 1.11: Collaborative competition between two or more TFs and a nucleosome can lead to cooperativity where initial binding of one TF weakens the interaction of the nucleosome to a patch of DNA and thus eases the access for another TF that binds to a site previously affected by the interaction [49].

has important functional implications during development [55, 56], where those are used as epigenetic switches, and in hormone-dependent cancers [14], since they respond to extracellular signals. For example, the pluripotency proteins Oct4, Sox2 and Klf4 were proposed to act as pioneer factors for c-Myc during reprogramming [57].

1.2.3 Complexation is important in development

Current genome-wide location analysis techniques like ChIP-seq (chromatin immunoprecipitation coupled with next-generation sequencing) enable to pin down locations and posttranslational modifications of proteins bound to the DNA. This is possible in vivo genome-wide with high sensitivity if appropriate antibodies for a specific target are available [58]. However, the specificity can be quite low, making it hard to differentiate between relevant and irrelevant binding events [51].

A variety of experiments examining colocalization of TFs, other regulatory proteins and chromatin states across all eukaryotes consolidate the significance of TF complexes as driving regulators in important regulatory pathways. The list of examples ranges from the cell cycle in yeast [59], body part formation in *Drosophila* and other model organisms [3, 36, 53] up to the comprehensive control of cell fate in mammalian stem cells [13, 51, 52, 60]. Grouped binding events show higher conservation between evolutionarily close species [36, 51–53] and have a greater impact on expression compared to individual binding events [52].

The developmental state of pluripotency in mammalian embryonic stem cells is controlled by a tightly interconnected network of several TFs [13]. Göke et al. [51] studied binding events of the key pluripotency factors Oct4, Sox2 and Nanog in embryonic stem cells in mouse and human. Between the two species less than 5% of individual binding events are conserved at orthologous genomic locations. However, about 15% of grouped events, where all three are factors are located within the range of protein interactions, are shared and at least 53% (Nanog, Oct4 even 63%) of such occurrences at developmental enhancers. This suggests a strict functional AND-relation mediated by protein-protein interactions. Additionally, colocalization of Mediator-subunits is significantly increased at those combinatorially bound conserved loci and correlates with an enrichment of distinct chromatin marks of activity (H3K27ac), clearly suggesting an effect of transcriptional activation [51].

Another related example from a review by Hochedlinger et al. [13] illustrates the previously stated importance of the final assembly: in embryonic stem cells colocalization of several pluripotency factors (among other TFs) often recruits coactivators such as the histone acetyltransferase (HAT) p300 to promotor regions of genes maintaining the pluripotent state, whereas individual binding of the factors results in transcriptional repression of genes associated with differentiation and lineage commitment by Polycomb group proteins (PcG) or histone deacetylases (HDACs) [13].

In summary, it can be confidently stated that TF complexes mediated by protein interactions are certainly of huge importance in essential cellular control circuits. As these recent publications show, a very nice property of such assemblies is that they might actually even hint at their functional mechanisms. Since we often have a certain knowledge of the catalytic activity of a recruited protein it can be possible to infer a regulatory effect from a given context.

1.3 Outline and Goal

TF complexes play important roles in the regulation of eukaryotic transcription in general and are increasingly considered as the driving regulators in crucial developmental pathways.
Without doubt, purposeful genome-wide localization experiments are the key to a profound understanding of the underlying regulatory networks and to reveal decisive new groupings of TFs, but the mechanics also suggest to take a look at a complementary idea: since multiprotein complexes are a foundation of promotor signal integration in eukaryotes, one could approach the problem in reverse direction and predict TF complexes based on data that is already available in abundance.

The knowledge of potential assembly candidates coupled with current data resources could provide the comprehensive information that is needed to infer possible target genes as well as the exerted mechanism of influence.
First of all, the TFs that are involved in the same complex together with their annotated binding sites should indicate which target genes to look after, since they are the DNA-binding "anchor"-elements according to our current models. Second, the further recruited proteins can show up the mode of action one can expect from such a complex.

The intended goal of this thesis is to develop and implement a novel approach that is able to predict such combinatorial assemblies of TFs.
For a first evaluation, yeast is the organism of choice since it is the standard in the assessment of complex prediction approaches, well annotated in general and the smallest eukaryote.

Related work

2.1 Protein complex prediction from networks

At a first glance, the prediction of protein complexes is a well-established problem. Data for physical interactions between proteins in the form of networks is quite abundant and so are clustering approaches that find dense areas in networks [1].

Clustering is a classical problem in computer science and mathematics and has the goal to group objects from a given universe into sets where objects within a set should be similar to each other and objects from different sets should be dissimilar according to an arbitrary similarity or distance measure [61, 62]. The exact definition depends on the specific goal and the context [62].
For a universe with a defined similarity measure one can easily transform the data to a graph structure by defining objects as the vertices and adding edges weighted by the measure for all pairwise comparisons. Additionally, edges could be discretized by removing the ones below a certain threshold and discarding the weights of the remaining ones. This yields an unweighted network in which only the strongly related nodes are connected [61]. The notion of a binary relatedness and its connection to distances enables to model even more problems within the framework of clustering, like social interactions or probable protein interactions. Finding the tighter related groups in a network, regardless of whether its edges are weighted or not, is then called graph clustering [61, 63, 64].

Protein-protein interaction networks (PPIN) offer a global view on the interactions between proteins of an organism by representing proteins as the nodes in the network and physical interactions, the protein-protein interactions (PPI), as an edge between the corresponding nodes. Plenty of methods exist to detect such physical interactions experimentally and modern high-throughput methods like yeast two-hybrid systems [65, 66] and (tandem) affinity purification accompanied with mass spectrometry [67, 68] made the annotation of whole organisms possible [69–72]. Unfortunately, the number of false positive and false negative interactions is very high. Indirect binding partners are, for example, often falsely reported as directly interacting ones. Furthermore, the overlap of datasets that are compiled using distinct experimental methods is surprisingly low due to individual strengths and weaknesses of each approach [70, 73–75]. The huge amount of uncertainty in the data is recently tackled quantitatively by the integration of different PPIN datasets and also additional heterogeneous data (like localization, coexpression or genomic evidence) into sophisticated statistical models. These models can then be used to obtain reliable weighted PPINs with interaction probabilities for each edge [69, 76, 77]. Section 3.2.2 will briefly address the Bayesian model that is used to construct PrePPI [71, 77], the PPIN used in this thesis.

Given that a set of proteins forms a complex to accomplish a common biological task, one expects this set to be a cohesive subset in such a PPIN [78–80].

According to [69] the complete characterization of a protein complex is a task that requires to solve several consecutive problems: (1) one needs to determine all member proteins of the complex, (2) the candidates must permit a connected topology of pairwise direct interactions, (3) these interactions have to be related to interactions between domains and distinct binding interfaces, and, with all this information one can (4) predict a 3D structure of the complex.

Section 2.2 will specifically address later steps, but most previous research has focused on the very first step which is basically a graph clustering problem. However, standard clustering is not ideal to detect complexes from PPIN. Many protein complexes are not only organized in a modular but also in a combinatorial fashion. They take part in several functional complexes and their associated nodes in the network therefore should belong to more than one cluster [81–86]. This has led to many sophisticated methods with very different underlying approaches. The first one, MCODE [80], works by iterative vertex reweighting, LCMA [87] searches for cliques and merges them, RNSC [88] optimizes a cost-function based on inter- and intra-complex edges, MCL [89] computes a flow within the network based on properties of its adjacency matrix, RRW [90] uses random walks in an iterative way and CFA [91] grows dense regions from k-connected subnetworks instead of cliques since protein complexes are not necessarily fully connected in a PPIN. A very successful recent method is ClusterONE [81] which will be explained in detail in Section 2.1.1.

Prediction methods always entail benchmarks to test their performance. Section 2.1.2 features common assessments in protein complex prediction that will be used in the thesis.
However, all the previously mentioned complex prediction approaches only apply graph-theoretic algorithms to the topology of the PPIN but neglect biologically important factors like structural limitations or combinatorial assembly which eventually leads to a high number of false positive predictions [69, 83, 86, 92–94]. Section 2.2 will elaborate the problems and introduce current solutions.

2.1.1 ClusterONE and cohesiveness

ClusterONE (short for clustering with overlapping neighborhood expansion) [81] is not only the opening track of Pink Floyd's 1994 album "The Division Bell"[1], but also one of the best performing complex prediction approaches available at the moment and one of the few that can handle weighted edges in PPINs as well as overlap of complexes.
It optimizes a very plausible metric for measuring the cluster quality called cohesiveness. The cohesiveness f for a set of proteins V in a network is defined as

$$f(V) = \frac{w^{in}(V)}{w^{in}(V) + w^{bound}(V) + p|V|}$$

where $w^{in}(V)$ denotes the total weight of edges between members of V (internal edges) and $w^{bound}(V)$ denotes the total weight of edges that connect V with the rest of the network (boundary edges). $p|V|$ serves as a penalty term with the purpose to model yet undiscovered interactions in the data. For $p > 0$ one assumes an additional boundary weight of p per protein in V. Figure 2.1 illustrates these definitions with an example and additionally introduces the notion of incident and boundary proteins.

Cohesiveness exactly assesses the structural properties of subnetworks we want to obtain: they should be densely connected but at the same time well separated from the outside. For example, $f(V) > 1/3$ implies that the proteins of the chosen subset have more internal than external weight on average. This satisfies the condition of a weak community [63].

Given some initial seed protein a growth algorithm iteratively increases

---

[1]http://en.wikipedia.org/wiki/Cluster_One

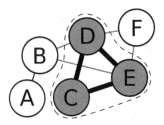

Figure 2.1: For convenience all edges in the network have unit weight and the corresponding weight annotation is omitted. The current members of the cohesive subset $V = \{C, D, E\}$ are shown darker and their internal edges are shown thicker, boundary edges are marked gray. Boundary edges can be thought to span a boundary (shown dashed) that separates the current dense subset V from the remaining network. This border defines the set of incident proteins $V_{inc} = \{B, F\}$, external vertices adjacent to the boundary, and boundary proteins $V_{bound} = \{D, E\}$, internal vertices at the boundary.
For the given example $w^{in}(V) = 3$, $w^{bound}(V) = 4$ and $f(V) = \frac{3}{7}$ if we neglect the penalty parameter p. Figure adapted from [1].

the cohesiveness using a greedy procedure. Based on the set of currently chosen proteins V, first the set of incident proteins V_{inc} and the set of boundary proteins V_{bound} are determined. All proteins in V_{inc} are adjacent to some protein in V and could be added to V in the step, all members in V_{bound} are on the boundary of V and could be removed from V. Figure 2.1 clarifies this on an example. For each of these possibilities to expand V or to shrink V, it is tested how the change would affect the cohesiveness $f(V')$ of the thus modified set V'. The algorithm then chooses the single addition/removal with the highest increase in cohesiveness as long as it can be further increased. If no further increase is possible it returns a locally optimal solution (see Algorithm 2.1).

ClusterONE can be run in a seeded mode, where the user explicitly supplies a list of seed proteins for its growth procedure algorithm, or one can leave this selection to the program. In this mode the growth is initially started from the protein with the largest number of connections (highest degree). After the completion of a single growth process, from all the proteins that are not yet included in one of the complex candidates so far, again the one with the highest degree is chosen as the next seed protein. This is done until there are no more proteins to consider.

In the next step of the algorithm highly overlapping complex candidates are merged. At first an overlap graph G is constructed where each previously

Algorithm 2.1 Iterative cohesiveness optimization starting in v_0

startprotein: $V_0 \leftarrow \{v_0\}$
$t \leftarrow 0$ {t: step number}
repeat
    $V_{t+1} \leftarrow V_t$
    determine current V_{inc} and V_{bound}

    check if addition/removal is valuable:
    for $\forall v \in V_{inc}$ **do**
        $V' \leftarrow V_t \cup \{v\}$
        if $f(V') > f(V_{t+1})$ **then**
            $V_{t+1} \leftarrow V'$
        end if
    end for
    for $\forall v \in V_{bound}$ **do**
        $V' \leftarrow V_t \setminus \{v\}$
        if $f(V') > f(V_{t+1})$ **then**
            $V_{t+1} \leftarrow V'$
        end if
    end for

until $V_t \leftarrow V_{t+1}$ {as long as further increase is possible}
return V {output locally optimal cohesive subset}

---

determined complex candidate is a vertex in the graph and two vertices A and B are connected by an edge if the overlap score $\omega(A, B)$ between the two sets is larger than 0.8 (for definition see Section 2.1.2). All candidates within the same connected component in G are then merged into single protein complex candidates; single vertices in G are carried over to the predicted output set without any merging step.

In the final step complex candidates with less than three members or below a certain density are discarded and the remaining ones returned.

2.1.2   Quality measures for complex prediction

Since complex prediction is already a well-established problem, plenty of benchmarks exist to assess the quality of such predictions. They can be clearly separated into two distinct but complementary categories: measurements based on the mutual agreement with reference complexes and measurements that account for plausible biological relationships within the predicted clusters.

Comparison with reference protein complexes

Given reliable and as complete as possible reference data of protein complexes for a certain organism, one can compare predictions to this compilation of known complex assemblies.

Unlike many other prediction problems, perfect matches to known complexes are rarely seen in protein complex prediction which raises the need for specifically adapted quality measures. A compilation of common ones is covered in the following paragraphs.

Overlap score and related scores

The overlap score suggested by Bader et al. [80] is a measure of overlap for pairs of sets and the foundation of many metrics to assess the quality of complex prediction approaches. Benchmarks based on the overlap score have been used in plenty of publications in slightly different variants and under different names [80, 81, 83, 93–95].

The overlap score ω between two sets of proteins A and B is defined as

$$\omega(A, B) = \frac{|A \cap B|^2}{|A||B|}.$$

Given some threshold t and $\omega(A, B) > t$ or $\omega(A, B) \geq t$ (depending on the exact definition) we call A and B matched.

The actual threshold can in principle be set as desired. In their original publication [80] evaluated this question for different thresholds and suggested t should be between 0.2 and 0.3 to get rid of biologically insignificant overlaps. In the remaining thesis we will use $\omega(A, B) > 0.25$ as used by [81] and [93]. The reasoning behind 0.25 is the fact that, given the compared complexes A and B are equally large, they share at least half of their proteins.

With the notion of a match one can compute a precision P as the fraction of predicted complexes that can be matched to a reference and a recall R as the fraction of reference complexes matched by predicted complexes. Additionally, one can define a combined F-score (or F-measure) using the harmonic mean

$$F = 2 \frac{PR}{P + R}$$

which is very common in information retrieval and was also used frequently in complex prediction [83, 95, 96]. All scores range from 0 to 1. $P = 1$ means that all predicted complexes can be related to true ones and $R = 1$ means that all known complexes are predicted.

The naming of the individual scores is inconsistent across publications but the quintessence is covered by this paragraph.

Geometric accuracy

The geometric accuracy goes back to Brohee and van Helden [92] and is based on a combination of the clustering-wise sensitivity Sn and positive predictive value PPV. Both metrics work on a contingency table $T = [t_{ij}]$, which is an $n \times m$ matrix where row i corresponds to the ith among the n reference complexes, column j to the jth predicted complex and t_{ij} denotes the number of shared proteins between reference i and prediction j. The cardinality of reference complex i is given by n_i. Then Sn and PPV are defined as:

$$Sn = \frac{\sum_{i=1}^{n} \max_{j=1}^{m} t_{ij}}{\sum_{i=1}^{n} n_i}$$

$$PPV = \frac{\sum_{j=1}^{m} \max_{i=1}^{n} t_{ij}}{\sum_{j=1}^{m} \sum_{i=1}^{n} t_{ij}}.$$

These two scores, each one optimizing a different direction, are joined using the geometric mean to obtain the final score:

$$Acc = \sqrt{Sn \times PPV}.$$

It is necessary to balance the two different metrics in the above manner since both can be artificially boosted by rather extreme predictions. The clustering-wise sensitivity Sn can be cheated by putting every protein in one huge cluster, while the positive predictive value PPV could be maximized by putting every protein in its own cluster.

The usage of the geometric mean instead of the arithmetic mean is beneficial in this case because then the combined score Acc is penalized stronger if one of the scores is comparably low. The arithmetic mean would give a false idea of the prediction quality in the extreme cases mentioned above: since one can easily get $Sn = 1$ or $PPV = 1$, an arithmetic accuracy would achieve $Acc_{arith} > 0.5$ for a hardcoded prediction [81, 95].

Maximum matching ratio

The Maximum matching ratio (MMR) is a rather recent quality measure proposed by [81] and is based on a maximal weight one-to-one mapping between reference and predicted complexes.

To establish this mapping a bipartite graph is constructed where reference complexes are related to predicted complexes by an edge if their overlap score is larger than zero. Now one can find one-to-one mappings using the definition of a matching from graph theory. A matching of a graph is a set of edges that do not share common endpoints, which means no prediction is assigned to several references and vice versa. A matching of a graph is maximal if there is no matching for this graph with a larger cardinality and maximum if there is no matching with a bigger sum of edge weights [97]. The MMR is defined as the sum of edge weights of the maximum matching divided by the number of reference complexes to be matched. The MMR is set in relation to the number of reference complexes and not offset against the number of predicted complexes. There are good reasons for this. Reference complexes are inherently incomplete, therefore unmatched predicted complexes should not be penalized [81].

Biological relevance

The second category of benchmarks is concerned with the assessment of the biological relevance of predicted protein complexes. It can be thought of as a biological plausibility check conceived to complement the measurements that rely on incomplete reference complex data.

Colocalization

The test for colocalization is based on the assumption that proteins within a complex must be locatable in a common compartment [76]. Given location data for all involved proteins, the colocalization score is defined as the average fraction of proteins encountered in the most common compartment within the complex weighted by the size of the complex [81, 98]. In the particular case of transcription factor complexes only the nucleus is an appropriate locality. Nucleus colocalization is thus defined as:

$$
Coloc_{nucl} = \frac{\displaystyle\sum_{\forall \text{ complexes } i} \frac{\text{members of complex } i \text{ located in the nucleus}}{|\text{complex } i|} |\text{complex } i|}{\displaystyle\sum_{\forall \text{ complexes } i} |\text{complex } i|}
$$

$$
= \frac{\displaystyle\sum_{\forall \text{ complexes } i} \text{members of complex } i \text{ located in the nucleus}}{\displaystyle\sum_{\forall \text{ complexes } i} |\text{complex } i|} .
$$

Gene Ontology enrichment

The last test has become a standard analysis in all areas of computational biology and assesses functional homogeneity within the complexes based on an extensive genome-wide annotation.

The Gene Ontology (GO) annotation [99] is a standardized representation for attributes of genes and gene products across species and databases. The GO defines a hierarchical relationship between annotation terms and is structured as a directed acyclic graph. Individual terms are represented as nodes and directed edges connect them to more specific terms to establish a parent/child-relationship where each term can be child of several parents. Consequently, for every term a gene is annotated with, it is furthermore associated with all the less specific parents of that term. The three distinct ontology domains, the roots of these trees, are:

molecular function: biochemical activity of the protein

biological process: biological objective to which the protein contributes

cellular component: localization where the protein is active

GO terms can be used to conduct an enrichment analysis. If a set of proteins is related functionally or members contribute to a common pathway one expects to find evident GO terms in the set more often than by mere chance. Thus, annotations that are found in an examined set of proteins are tested for overrepresentation against a suitable background such as all genes of an organism or all genes covered by the microarray used in the study. This type of analysis has become a quasi-standard in the investigation of biological data and was also used in the context of protein complexes before [81, 95, 100, 101].

The probability P to observe k or more proteins annotated with term X in a set of m proteins by chance is given by a hypergeometric distribution:

$$P = \sum_{i=k}^{m} \frac{\binom{M-K}{m-i}\binom{K}{i}}{\binom{M}{m}}$$

where M denotes the number of proteins in the background, m the number of proteins in the studied set, K all proteins in the background annotated with term X, and k the number of proteins in the studied set annotated with X. Term X is then said to be enriched in the studied set at significance level p if $P < p$. Since multiple hypothesis testing is performed to examine if the set of interest contains an enriched GO term, significance levels have to be adjusted. A simple and conservative method to correct them is the Bonferroni correction whereby the significance levels are simply divided by the number of tests [102].

To utilize the idea of GO enrichment for the assessment of complex prediction quality the overrepresentation score is defined as the fraction of complex candidates with at least one enriched annotation at significance level $p = 0.05$ [81, 100].

2.2 Protein complex prediction beyond protein interaction networks

PPINs provide a holistic view on protein connectivity which, with respect to applicability in the prediction of protein complexes, misses certain important layers of information [69]. Interaction networks offer a compilation of assumed pairwise interactions that are thrown together to a static entity but in reality the network is highly dynamic and intrinsically controlled by protein expression, affecting the current state of the network in time and space, and spatial constraints. To enable complex formation all involved proteins must be expressed at the same time, in spatial proximity and capable of forming a stable binding topology devoid of any binding site competition [69, 83–86, 103, 104]. Hence, predicted clusters in PPINs are not necessarily valid biological complexes and their interpretation can be quite ambiguous and even lead to false positive complex predictions as Figure 2.2 illustrates. Nonetheless, dense connectivity and slight seclusion still suggest that clusters in PPINs form at least functional modules, what are groups of proteins transiently interacting with each other on cellular functions as overlapping complexes [78, 82, 83, 86]. The integration of additional heterogeneous data allows to derive a clearer picture.

Figure 2.2: Fully connected subgraphs are the densest subsets one can find in a graph. Unfortunately, in the biological context it is not clear how to interpret such an clustering result. From the sole connectivity it is not possible to properly distinguish between this being a complete protein complex or just the outcome of several pairwise interactions $\{A, B\}$, $\{A, C\}$, and $\{B, C\}$ or even less interactions and false positives.

Different studies combined interaction and expression data of yeast to reveal clear evidence for such a dynamically organized modularity in PPIN [84, 103]. When Han et al. [84] examined expression profiles among the interaction partners of hubs, namely highly connected proteins in the network, they could clearly distinguish between two basic types: 'party hubs', whereby all partners and the hub protein are considerably coexpressed, and 'date hubs', where this is not the case. Rigorous analysis clarified the functional purposes and the biological meaning behind this dichotomy. The rigid coexpression in party hubs is needed to establish fixed assemblies of complexes or functional modules, subnetworks characterizing the 'lower level' buried inside functional modules. In date hubs interactions with the same protein often occur at different times or localizations demonstrating combinatorial complex assembly in vivo. They are considered as the 'higher level' connectors between modules [84]. For the scope of the thesis this means that different TFs and TF complexes as well as coregulators can be combined in this way.

The next question one could ask is, given all proteins were together at the same time and in the same place, which interactions can actually occur together at all and thus separate real complexes from functional modules? Kim et al. [85] tackled this question by building a subnetwork of a yeast protein interaction network consisting only of proteins with known structure and edges matching known homologue interface interactions. The resulting subnetwork not even contained 15% of all yeast proteins (873 proteins

of around 6000 [70, 76, 80]) and an even smaller fraction of interactions
(1269 interactions, estimated lower bound 30, 000 [70, 76, 80]). Despite the
relative incompleteness, the construction already showed that dynamic of
protein interactions also happens on the level of binding interfaces since
a third of all interactions excluded each other (438 of 1269 were mutually
exclusive). Conversely, this means that there are many possible coactive
states of the same network that include only a subset of the interactions,
but are free from binding interface competitions.

Following these discoveries Jung et al. [83] investigated the usage of such
simultaneous protein interactions networks (SPIN) in the prediction of protein complexes.
They took a protein interaction network of yeast and marked mutual exclusiveness where data from a structural interactome map by [105] suggested
so. Parts of the network without any structural annotation were taken as
fixed, in contrast to [85] no network data was omitted in their approach.
To obtain subnetworks without any conflicts one needs to choose from edges
that are in conflict with each other and enumerate all valid edge-decisions.
Because every edge involved in a conflict can either be existent or not,
exponentially many SPINs are possible. Since this is not feasible for the
complete network of yeast they first determined dense areas using conventional clustering approaches and restricted the construction of SPINs to
the pre-processed parts. These were used to generate the simultaneously
possible subnetworks, which then are finally used to predict complexes.
To our knowledge, this is the only current method that accounts for the
combinatorial possibilities due to binding interface limitations. The au-
thors compared the approach to conventional clustering methods and to
randomly constructed SPINs to investigate if possible improvements are
based on the added structural information. The evaluation showed that
the effort of SPIN construction in many cases leads to a considerable refinement of a given pre-clustering by the consequent exclusion of superfluous
proteins from complex candidates and the more reliable discovery of overlapping complexes that share common core proteins but involve exchange-
able adaptor subunits. A common complex prediction algorithm, even if it
supports overlaps, might not be able to dissect such highly modular complexes, but the additional analysis of possible binding site competitors and
the enumeration stage reveals the combinatorial possibilities.
However, the benefit of the approach is limited by the amount of available
structural data, which in turn is still insufficient. Furthermore, the combi-

natorial complexity of the SPIN construction enforces a limitation to dense areas in the network (pre-clustering) and the approach has no notion of density itself (post-clustering) which leads to an inevitable dependency on conventional methods. The choice of the algorithm(s) has a huge impact on the final results.

2.2.1 Domain-domain interaction model

Due to the sparsity of structural data, a nice new model for the question of concurrent connectivity came up: the domain-domain interaction model by Ozawa et al. [93]. Given a dense protein-protein interaction subnetwork from a generic complex prediction approach, one decomposes the proteins into their domains and justifies connectivity on interactions between the individual domains, the domain-domain interactions (DDI). This transformation to a 'domain-level' network can then be used to filter out false positive predictions. If every domain is restricted to participate in only one of those interactions, the model implies a certain binding interface constraint and thus reveals which proteins can be connected at the same time. This is important because true protein complexes must admit a topology of pairwise binding on the level of interfaces and these interfaces are often exclusive [69, 85, 86, 106, 107]. A protein complex predicted by clustering on a PPIN that is said to be valid according to the model therefore has to allow a choice of DDIs that leads to a connected component in the domain-domain interaction network (DDIN). Although there are occasional instances of single domains that are possibly bound by multiple domains concurrently, in the interest of simplification the model equates individual interfaces with single domains [93, 94, 108]. The big advantage of a model based on domain annotations is the wealth of established interference methods and well-stocked databases that are able to unravel the domain composition of a protein using diverse data [109–112].

The approach of Ozawa et al. is formalized as a problem in binary integer linear programming [113], where all variables are either zero or one. In the formal definition by [93] Ω denotes the subset of all protein interactions in a candidate complex from the original PPIN, $P_{i,j} \in \Omega$ is the potential PPI between proteins p_i and p_j where $P_{i,j} = 1$ denotes the case of an established interaction. Their objective is to maximize the number of simultaneous protein interactions within the candidate:

$$\text{maximize} \sum_{P_{i,j} \in \Omega} P_{i,j}$$

A PPI has to be established by an associated interaction between domains. Let $D_{i,j} = \{D_{i,j,1}, D_{i,j,2}, \ldots\}$ be the set of potential domain interactions that could connect p_i and p_j. A protein interaction variable $P_{i,j}$ becomes 1 if any domain interaction in $D_{i,j}$ is set to 1:

$$P_{i,j} = \sum_{D_{i,j} \in D_{i,j,k}} D_{i,j,k}$$

According to the assumptions of the model, every domain should only be able to participate in at most one DDI. For a domain d_m within a protein of the candidate complex $D(d_m)$ is the set of all DDIs where d_m is involved. Then the domain-constraint can be formulated in ILP as:

$$\sum_{D_{i,j,k} \in D(d_m)} D_{i,j,k} \leq 1$$

Every connected component of the resulting subnetwork containing more than two proteins and more than one domain interaction is then assumed to be a valid protein complex. Figure 2.3 illustrates the principle with an example.

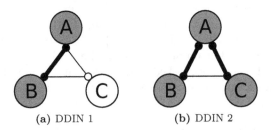

(a) DDIN 1   (b) DDIN 2

Figure 2.3: Subfigures **(a)** and **(b)** show different examples how a DDIN of the PPIN in Figure 2.2, a clique of three nodes, could look like on the level of domains. While **(a)** does not admit a connected component $\{A, B, C\}$ within the constraints of the model, the domain-architecture in **(b)** allows the complete complex candidate. As a side note, **(a)** actually has three equivalent solutions: $\{A, B\}$, $\{A, C\}$ and $\{B, C\}$. Neither ILP nor maximum matching have a notion for equivalent solutions, only one is reported.

In a further publication another group proposed some extensions to the method and improved running time and sensitivity by using maximum matching instead of binary integer linear programming and by adding artificial interactions [94].

The original optimization problem aims to maximize the number of edges in the PPI graph while each vertex in the DDI graph (each domain) is constraint to participate in at most one interaction. Ma et al. [94] maximized the number of DDIs directly while conserving the same constraint. This goal exactly resembles the maximum matching problem which aims to find a subset of edges MM in a graph G with particular properties [97]. Unlike the binary integer linear programming formulation, an approach based on maximum matching can be solved in polynomial time for any graph G [97, 114].

Missing data is also a problem on the domain-level. Many proteins have incomplete domain annotation and very reliable domain-domain interactions are infrequent [94, 111, 112, 115]. When Ma et al. [94] dissected their predictions they found many examples that lacked proteins in comparison to the associated reference complexes although the missing proteins reside in close neighborhood to predicted candidates in the PPIN. All these proteins missed DDIs to support the PPIs to their dense periphery and were therefore isolated. Being aware of these deficiencies, they improved the recall of their approach by adding artificial interactions. If an isolated protein has at most one domain and none of its neighbors in the PPIN that form a valid complex according to the model has more than one annotated domain, then the isolated protein is randomly connected to one of these neighbors. As a result, fewer proteins are lost due to missing connectivity data.

Deploying the domain-domain interaction model as a filtering step to existing clustering approaches increases the accuracy of predictions by the elaborate introduction of biological constraints [93, 94]. Unfortunately, the choice of the pre-clustering method has a decisive impact on the final result leading to a strong bias. However, the critical flaw for our purpose is the absence of alternative solutions since ILP and maximum matching can only give single solutions to an optimization problem (see Figure 2.3). If one is interested in the combinatorial manifold of considerable solutions - like us - one would still have to enumerate all possible exclusive choices.

2.3   Prediction of cooperative TFs

The importance of cooperativity between TFs was already clarified in Section 1.2.2.1. However, cooperativity in a general sense is not well-defined. The aim to predict potentially cooperative pairs of TFs yielded diverse methods which have mostly been applied to yeast and to the yeast cell cycle, if expression data was incorporated [48, 50, 116–120]. However, even when the predictions are comparable, the overlap of their individual results is rather low, yet significant [48].

Some methods are based on noticeable binding site cooccurrence in the promotors of target genes [50, 121]. Moreover, dynamic models to detect non-linear behaviour in expression time-series have been employed [117] as well as statistical tests, whereby genes are discretized according to their expression in certain phases of the cell cycle [120].

Other methods are based on the idea that a certain measure for gene sets changes significantly if the constrained set targeted by a decisive pair is compared to the target genes of the individual TFs. Such measures include the average distance within the PPIN [119], are grounded on the assumptions of proximity among proteins that work together and should therefore be expressed together, or exploit the expression coherence [116, 118], a measure of coexpression which should be higher among genes controlled by the same mechanism.

In contrast to some of the specialized approaches, expression coherence is a very general concept allowing to integrate different types of expression data and therefore valuable to be further exploited.

2.3.1   Expression coherence scoring

Expression coherence (EC) is a measure of coexpression for sets of genes. The basic idea is to combine all pairwise distances between the expression profiles of the genes in a way to define a score for the whole set. Which distance measure is exactly used and how the pairwise distances are combined is not strictly specified [116, 118, 122, 123].

Common distance measures of coexpression are the euclidean distance, the correlation coefficient (as a distance) or more sophisticated (but not necessarily superior) metrics like the mutual information [124]. The pairwise information can be integrated to a score in several ways, for example as the average of pairwise gene expression profiles or the significance of such an average against the mean and standard deviation of random genes [122].

For examined sets of expression profiles that are split into a few very tightly clustered subsets a score calculated in one of the above ways could still be low. Thus Pilpel et al. [116] proposed another measure which became the standard to define a scoring for a set of genes. They defined the expression coherence score (ECS) as the fraction of pairwise interactions below a certain threshold D, where D is determined as the 5th percentile across the pairwise distances between 100 randomly sampled profiles from the whole data set. Figure 2.4 visualizes the crucial difference this property introduces in theory when it comes to clustered sets and Figure 2.5 in practice, showing the individual expression profiles of all genes from two similar sized gene sets that feature very different ECSs and feature clustered subsets.

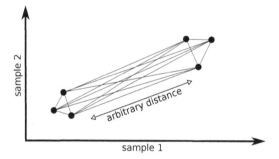

Figure 2.4: A coexpression measure like the average pairwise distance in a set of genes is quite susceptible if densely clustered subsets are remote to each other. A measurement based on the fraction of tighter connections, on the other hand, is unperturbed by the inter-cluster distance.

While the original publication uses the euclidean distance to measure pairwise coexpression, the correlation formulated as a distance has become the prevalent standard [118, 121, 124].

Significance of expression coherence scores and their change

Although the ECS itself is a very comprehensible measure, the mere score has little meaning on its own since it is strongly dependent on the size of the evaluated gene set. Hence, a simple p-value was suggested on the hypothesis that a given ECS for a set of genes from a certain expression profile can be observed by randomly drawing gene sets of equal size from the same dataset [123]. The approximated p-value of the examined gene set is then simply the fraction of sets chosen at random, but with equal

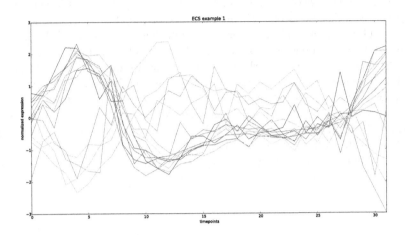

(a) high expression coherence score: about 0.23

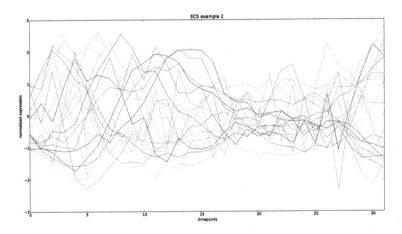

(b) low expression coherence score: about 0.02

Figure 2.5: Both plots show time-series gene expression profiles from the yeast cell cycle for common target genes of distinct pairs of TFs. While the genes in **(b)** do not seem to be related, manifesting in a low ECS, the genes in **(a)** can be clearly clustered into two subsets which behave quite contrary. The profiles and EC scores are obtained using the data and methodology as presented in the Methods section.

size and dataset as the set to be tested, with at least the same ECS. The ECS of example **(a)** in Figure 2.5 is significant.

In Pilpel et al.'s cooperativity prediction approach a pair of TFs A and B is said to work synergistically if the ECS of the shared target genes (set AB) is significantly greater than that of targets bound by only one of the two (sets A and B). This was tested for all distinct pairs of yeast TFs for cell cycle expression data and for different conditions [116].

Their significance estimation is also based on a Monte Carlo procedure, but this time the change of the ECS is sampled. Gene sets AB and $A \setminus B$ are a partition of TF A's target genes A distinguished by their ability to be bound by TF B or not. Assuming gene set $A \setminus B$ (all genes bound by TF A, but not B) has a higher ECS than $B \setminus A$, the induced coherence change by the constraint of the targets dECS = ECS(AB) - ECS($A \setminus B$) is the smaller of the two. Now to test the null hypothesis that ECS(AB) is less than or equal to ECS($A \setminus B$) dECS is compared to the ECS differences obtained by random partitioning. In every sample step A is split into two sets S_1 and S_2 of size $|AB|$ and $|A \setminus B|$ at random, and ECS(S_1) - ECS(S_2) is computed. Repeated T times, a distribution of ECS differences is obtained that can be used to calculate a significance of dECS as the fraction of random differences that are larger or equal to dECS, with $1/T$ as the upper bound. To avoid overestimation of individual pairs, the authors suggest to set T to the number of generatable hypotheses at most, thus in their case the number of TFs squared [116].

Materials and Methods

The method proposed in this thesis utilizes ideas from some of the previously mentioned publications and combines worthwhile features to a novel approach adjusted to our goal of the identification of transcription factor complexes.

The algorithm uses the DDI-model (see Section 2.2.1) to account for the exclusive and thus combinatorial nature of certain interactions but at the same time optimizes the cohesiveness (see Section 2.1.1) on the more holistic and better resolved level of protein interactions. Since we are only interested in complexes of TFs such an algorithm can be formulated in a local way which makes it possible to run it on the unbiased complete network in a feasible way.

The first section aims to deliver the overall idea of the method without getting lost in any technical details, yet it should provide a clear understanding of a naive implementation and the data that is integrated and therefore a strict prerequisite.

Subsequently, all data sources that are used within the thesis will be introduced and justified. Data retrieval procedures and preprocessing steps are included where necessary.

Section 3.3 will finally cover the actual implementation in detail and address two conceivable algorithmic variants.

3.1 Domain-aware cohesiveness optimization

As already mentioned, the algorithm does iterative optimization of the cohesiveness up to a local maximum as proposed by [81] but constrains the degrees of freedom within the network using the DDI-model to approximate binding interface competition. The affiliation of these slightly different levels of granularity coupled with the set of rules introduced by the domain model enforces a loop invariant within the main algorithm that ensures a conflict-free interaction topology between the currently grown dense subset during the whole execution and thus also at termination. Hence, the cohesive local choice of proteins is always accompanied with a valid spanning tree on the domain level.

Given a certain current execution state of the algorithm, for simplicity defined as the set of currently chosen proteins in the assumed cluster V and the domain interactions between them, the first thing that is done in each step is to determine incident and boundary proteins, just like in ClusterONE. But in our approach an additional network level is introduced: connectivity questions are answered based on the topology of the domain-domain interaction network and the constraints of the domain-domain interaction model (Section 2.2.1). With the introduction of this abstraction, binding site conflicts as defined by the model are precluded right from the start to retain the self-imposed loop invariant. Figure 3.1 illustrates the principle with an example and thereby introduces the refined definitions for incident and boundary proteins.

In the next step, it is checked for all incident proteins how valuable it is to add them in terms of cohesiveness $f(V')$ and for all boundary proteins how valuable it is to remove them. This computation is solely done on the level of the abundant and carefully weighted protein interactions solely and is independent of the current domain occupancy in the sense that all interactions are taken into account. The advantage of using the data unrestrictedly is a certain noise-cancellation effect. Individual edge weights may be unreliable, but when many interactions are combined slight deviations can be expected to average out. From all possible modification V' the cohesiveness-maximizing one is chosen.

Every iteration can have three outcomes: the algorithm could terminate since no further increase is possible, the removal of a protein could yield the largest gain or the addition of one may be the most favorable choice.

In the first case the current complex candidate set V is returned and the algorithm terminates, since the cohesiveness is locally optimal.

The removal case changes the current state. The boundary protein P is removed from V leading to $V' = V \setminus \{P\}$ and additionally the domains occupied by the distinct spanning-edge that connected P to the remaining cluster V' are regenerated. The definition of boundary proteins ensures the preservation of connectivity within domains of V'. The next iteration is then computed using this modified state.

The algorithm becomes time consuming when a protein P should be added. Naturally, V is modified to $V' = V \cup \{P\}$ and P needs to establish a domain interaction with a protein from V to maintain internal persistence. Often several interactions on the domain-level are able to link P to V on the

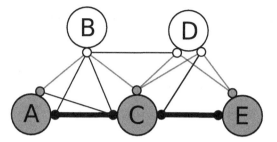

Figure 3.1: In this figure of an DDIN the gray nodes are the proteins currently in $V = \{A, C, E\}$ and the thick edges show how the current dense cluster is connected on the domain-level. Incident nodes $V_{inc} = \{B, D\}$ are the ones that can be connected to V by a new domain-interaction edge which needs an unused domain of an internal protein (gray smaller nodes, the domains). Boundary nodes $V_{bound} = \{A, E\}$ are proteins in V with only one used domain. A later removal of an internal node with two or more occupied domains would inevitably lead to a breakdown of the spanning tree on the domain level and thus introduce inconsistencies during execution. Figure adapted from [1].

domain level. This edge choice can be crucial for further expansions since it leads to differing occupations on the domain-level and therefore changes the moves that will be possible in later steps. However, at this network level there exists no intrinsic rating of better or worse for each edge since all choices lead to the same increase in the function that is optimized (at least not without further assumptions, more on this in Section 3.3.2.2). To take these choices into account, the algorithm branches and needs to evaluate the outcome of every spanning edge on the domain level.

Algorithm 3.3 summarizes the key steps during one iteration, the iterative part will be covered later.

A staged example will illustrate how this approach can lead to a set of distinct solutions from a common starting point if such an outcome is conceivable. Figure 3.2 shows the relevant snapshots of the subnetworks for the two different levels of granularity that are used by the method. For every step, first the DDIN in its current execution state is shown, resolving the question of constraint reachability, and afterwards the resulting best choice of proteins V' in terms of the cohesiveness within the PPIN. A stepwise explanation of the process follows directly after some remarks about the information within the graphics.

Each edge in the PPIN is weighted by its reliability, dotted lines with weight

---

Algorithm 3.1 Basic domain-aware cohesiveness optimization step function:

$\text{step}(V, E)$

---

given current proteins V and used domain interactions E {as described later, actual state description can differ}

determine V_{inc} and V_{bound} from the DDIN and the used edges E
$max \leftarrow f(V)$
$action \leftarrow$ terminate

for $\forall P \in V_{inc}$ **do**
    $V' \leftarrow V \cup \{P\}$
    if $f(V') > max$ **then**
        $max \leftarrow f(V')$
        $action \leftarrow$ add P
    end if
end for
for $\forall P \in V_{bound}$ **do**
    $V' \leftarrow V \setminus \{P\}$
    if $f(V') > max$ **then**
        $max \leftarrow f(V')$
        $action \leftarrow$ remove P
    end if
end for

if $action =$ terminate **then**
    return result V {termination yields a complex candidate}
else if $action =$ remove P **then**
    $V' \leftarrow V \setminus \{P\}$
    determine $e \in E$ that connects P to V' in DDIN
    return job $(V', E \setminus \{e\})$ {removal yields the state description for the next step}
else if $action =$ add P **then**
    $V' \leftarrow V \cup \{P\}$
    initialize empty list l
    for $\forall e \in$ DDIN so that $E \cup e$ spans V' **do**
        append $(V', E \cup e)$ to l
    end for
    return list of jobs l {addition yields a list of state descriptions that need to be further processed}
end if

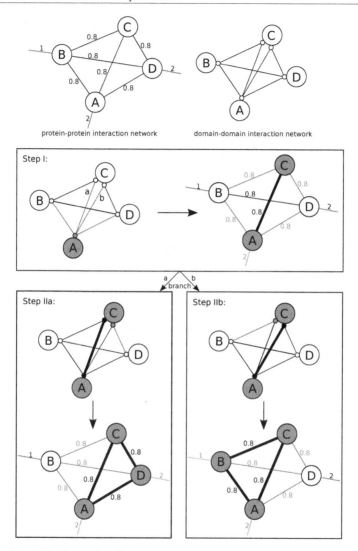

Figure 3.2: This illustration shows the key steps and network states that are traversed during the execution of the proposed algorithm on an example. Due to lack of space, the detailed step-by-step explanation is given in the accompanying text. Figure adapted from [1].

annotation signify the sum of weights from additional outgoing interactions that are, for the sake of convenience, not shown in the example. Like in PPINs before, internal edges are shown thicker and current proteins in V are colored darker, boundary edges are shown in gray. The corresponding weights are colored accordingly. Additionally, in the depicted DDIN snippets usable interactions are marked gray and established ones are made thicker, occupied domains are colored black and free domains of proteins in V highlighted gray to mark availability for enlargement. The domain architecture is intentionally chosen to suggest a certain dynamic modularity that cannot be inferred from sole protein interaction data, as presented in Figure 2.2 and the corresponding section. The growth process is started from protein A, thus V is initialized as $V = \{A\}$.

Step I: A's single domain is currently not occupied, so all its domain interactions are usable to extend the complex candidate (possibilities are highlighted in red). Reachable targets on the domain level are B, C and D. For all those possibilities the cohesiveness in union with A is computed on the PPI-level. The largest gain is obtained by the addition of C, thus C is added to V and $V' = V \cup \{C\} = \{A, C\}$. The current state of the PPIN is annotated accordingly. To preserve the consistency of our model, V' needs to be backed by a spanning tree of domain interactions. The tree can be initialized to include C by two equivalent choices of domain interactions, highlighted with **a** and **b**. Since both possibilities are conceivable, the algorithm branches into two alternative scenarios that consider the individual consequences independently.

Step IIa: In this branch domain interaction **a** is chosen to establish a connection between A and C, thus A's domain and C's left domain become occupied. C's right domain is the only domain of an internal proteins that can support an additional connection. Therefore only an enlargement of the complex through the inclusion of D is possible in the thus constraint DDIN. For complex candidates of size two a removal is never worthwhile and thus not checked. Since adding D increases the cohesiveness and there is only one distinct domain interaction enabling this extension, the modification is performed and no branching is necessary. Since no further increase is possible, $\{A, C, D\}$ is returned by the branch.

Step IIb: In this branch domain interaction **b** is chosen and only C's left domain can still be used for expansion. The situation is symmetric to Step IIa and the details are omitted. The execution terminates with the output $\{A, B, C\}$.

Result: The approach detects two distinct complex candidates $\{A, C, D\}$ and $\{A, B, C\}$ from a common start. Generic cohesiveness optimization that only uses data of protein interactions would have returned the whole protein set $\{A, B, C, D\}$, neglecting the underlying combinatorial possibilities and interface-limitations which are here accounted for by the inclusion of the DDI-model. A similar example can also be found in the supplement of [1].

The approach has certain prominent amenities. As shown by ClusterONE [81], cohesiveness is an excellent-performing cluster-quality measure on its own and the additionally employed domain-decomposition allows to differentiate between interactions that are possible simultaneously and those that are not. The way the procedure works, the branching of distinct spanning-tree realizations on the domain level, naturally leads to several solutions for a common start if topology suggests so. Since we are only interested in complex formation of TFs we can focus the growth around them and thus keep the combinatorial explosion on a local level, something that was not considered in any of the approaches presented so far (see Section 2.2). Since the computational cost is still high, algorithmic optimization is certainly a worthwhile effort.

3.1.1 Why seeding from pairs is beneficial

Local cohesiveness optimization as described by Nepusz et al. [81] starts its greedy growth process from single proteins in the network. Like every greedy algorithm, it proceeds shortsighted and is thus prone to local extrema [97, 125]. In general this behaviour is, at least to some extent, even desirable for our task since we want to grasp the combinatorial manifold of complexes only locally around TFs and, due to the implemented branching principle, the algorithm is even able to diversify its intermediate states with a variety of justified directionalities induced by the previous history of domain choices.

However, each time the complex candidate is enlarged only one protein is chosen for expansion, regardless of how big the differences between individual choices are. This is especially critical and error-prone during the very first expansion-step for which only one internal edge and the boundary edges of the two proteins from our inherently noisy network data [70] are considered for score computation. A preferential bias at this stage is

dangerous for our definite scope because it precludes the opportunity to discover reasonable combinatorial alternatives in advance. Without any previous history, local 'valleys' (more descriptive, for cohesiveness hills would actually be correct) become final whereabouts and cannot necessarily be escaped by branching later on.
Figure 3.3 illustrates the problem in detail on a task-specific example.

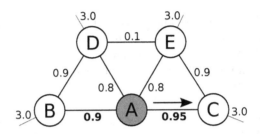

Figure 3.3: This example shows a snippet of a PPIN where each edge is weighted with an estimated probability of interaction and dotted lines sum up additional outgoing weights of interactions that are not shown. In our context, protein A could be a known TF and the other proteins potential coregulators. The subgraph topology and the weights per edge are selected reasonably to resemble potential biologically meaningful complexes $\{A, B, D\}$ and $\{A, C, E\}$ that should be predicted. The network is almost perfectly symmetric besides two weights (shown bold) that deviate within a scale we should at least expect from noise of corresponding data generating methods and reliability predictions in practice. Such tiny perturbations are sufficient to bias the complex prediction based on local cohesiveness optimization exclusively to the right half of the pictured network. If one executes the algorithm by hand starting in A, the prediction will first add C and then become trapped in the local maximum $\{A, C, E\}$. $\{A, B, D\}$ cannot be reached, even if the domain information is incorporated. An algorithm initiated like that can become stuck early and combinatorial details could be lost irrevocably in advance. Figure adapted from [1].

Considering this pitfall, it seems reasonable to start the growth process from pre-built pairs of proteins to spice up the optimization. This would overcome any unfounded directional bias early on in the first expansion step and pave the way for the reasoned bias due to domain constraints. Furthermore, the pairings should be determined on the basis of the probability of a protein interaction between the two proteins up to a certain likelihood if such data is available.
The rationale behind this decision is that while cohesiveness (see definition in Section 2.1.1) is undoubtedly an expressive measure of cluster quality if several proteins are involved, the included notion of seclusiveness only has

a limited validity for pairs of proteins. It is especially misleading for proteins that are combinatorically active and thus potentially exhibit higher boundary weights within the cohesiveness calculations putting them to a disadvantage compared to less promiscuous nodes in the network.

At this point one could argue whether the tedious inclusion of the domain model is really a necessity to capture the combinatorial freedom of protein complexes. It seems as if one could just simply modify an existing local cohesiveness optimization method like ClusterONE [81] to start from such curated pairs instead of single proteins. This is a question worth to evaluate on real data and will be addressed in Section 4.2.
While this would actually work for the exact example in Figure 3.3, because it has its 'preferable valleys' directly adjacent to the iteration-start and they will trap the algorithm, it would fail if the outer weights (dotted edges) were lower (say 1.0) and thus allow for a growth over the whole shown subnetwork missing the individual units. With the addition of the domain constraints, all those cases are covered reliably. Figure 3.4 summarizes the suggested scheme with the aid of a hypothetical cohesiveness landscape.

3.2 Data sources, their retrieval and preprocessing

3.2.1 General yeast protein data

The process of data retrieval is started with the download of a list containing all entries for yeast (*Saccharomyces cerevisiae*, taxon identifier 4932) in the Universal Protein Resource database (UniProt) [126, 127], a comprehensive catalogue for protein sequences and a wealth of annotation data that is updated monthly with the latest information from all integrated external databases.

The up-to-date list for yeast can be found in the documentation of the current release on ftp://ftp.uniprot.org/. The December 2013 release comprises 6624 proteins. UniProt IDs (or UniProt accession numbers) are strings of six alphanumerical characters that are used within the scripts to name proteins unambiguously. "P25302" is the identifier for the regulatory protein SWI4, for example. From UniProt we retrieved the following data for each protein: its mass, genomic locus, corresponding open reading frame

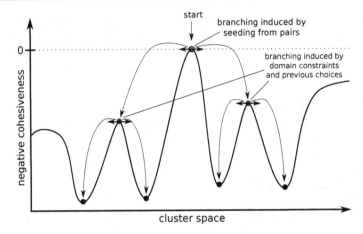

Figure 3.4: Given its current state, every step in local cohesiveness optimization can only accomplish a certain improvement (dotted arrows). This is done iteratively until a local optimum is reached (black points on lowest level). As already explained in detail, one possible transformation, the addition of a protein to the current cluster, can lead to alternative branches of domain choices. Such previous history can induce justified local biases grounded on the assumptions of the DDI-model (gray points). With the additional introduction of pairwise seeding a better coverage of the sampling is ensured and unfounded local traps are avoided (starting point). An efficient implementation will later on take care of pointless additional algorithmic overhead.

(ORF), name(s) and synonyms, the length of its amino acid sequence, associated InterPro domains [128] and finally GO annotation terms. Only 16% of the proteins are structurally resolved and have associated entries in the Protein Data Bank (PDB). The bundled naming information therein and additionally the UniProt ID mapping web service are used for later name conversion tasks.

All entries for yeast were preloaded and saved in a serialized binary format for further usage using Python's 'cPickle' module [129]. Each time the UniProt-handler class is imported for usage, the serialized data is read and internally made available as a fast map data structure. This avoids unnecessary waiting times during the execution of the main algorithm to fetch information from web services. Additionally, every entry is tagged with a version-field to preserve an internally consistent data structure. Such caching and serialized saving to disk for later usage is done consequently for all data sources from web services and omitted in further explanations.

3.2.2 Weighted protein-protein interaction data

PrePPI [71, 77] combines the interactions reported by six experimental yeast protein-interaction datasets for yeast (MIPS [130], DIP [131], IntAct [132], MINT [133], HPRD [134] and BioGRID [135]) with novel predicted interactions and employs a Bayesian framework to assign to each edge in the constructed network a probability of being a correct true-positive interaction.

The reliability of their prediction approach based on structural data is at least comparable to high-throughput experiments [77]. The other big advantage of PrePPI is the estimation of probabilities for each interaction. The basic principle was introduced in an earlier publication by Jansen et al. [76]: based on available non-structural data for all proteins and reference sets for gold-standard positive interactions, compiled from various literature, and gold-standard negatives, built under the assumption that proteins in different cellular compartments never interact in vivo, likelihood ratios can be computed in a training phase and then used to calculate probabilities. The likelihood ratio $LR(f)$ for one discrete feature f is then given as

$$LR(f) = \frac{\text{fraction of positive reference with feature } f}{\text{fraction of negative reference with feature } f}.$$

Assuming uncorrelated clues, likelihood ratios for several features f_i can be combined by taking the product

$$LR(f_1, f_2, \ldots, f_n) = \prod_{i=1}^{n} LR(f_i)$$

and the final probability is then given as

$$P = \frac{LR}{LR + LR_{\text{cutoff}}}$$

whereby a prior $LR_{\text{cutoff}} = 600$ is introduced to shift a LR of 600 to a probability of 0.5 of being a true interaction. The prior is in principle an arbitrary assumption about the true number of positives. Based on previous estimates it is conservatively suggested that on average there is one

true PPI among 600 protein pairs [76, 77, 136, 137]. Features that are
combined within the model are the essentiality of protein pairs (are the
proteins essential for the cell or not), coexpression, functional similarity
regarding GO and MIPS, and evolutionary similarity.

The PrePPI database only allows to query single proteins by name or
UniProt ID and thereupon returns possible interaction partners together
with the corresponding predicted probabilities of a direct interaction between each partner and the requested protein. To obtain the entire network,
the previously downloaded list of all UniProt proteins was used as an initial
seed for a special web crawler script that gathered all links and probabilities
by a breadth-first search [97] on the expanded queries. The thus received
data was then exported to a file in the convenient Simple Interaction Format (SIF), which is understood by all tools that deal with graph-like data
such as clustering programs or the popular biological network analysis and
visualization platform Cytoscape [138].

The complete PrePPI protein-protein network for yeast (version 1.2, released Jan. 2012) consists of 6198 proteins and $235,527$ interactions. Restricted to interactions with $P \geq 0.5$ the network still has $61,720$ edges.

3.2.3 Domain-association and domain-interactions

To build a domain-domain interaction network one needs to be able to
subdivide proteins into their domains and to have a reliable collection of
interactions between those domains.

3.2.3.1 Associating domains with proteins

First, proteins need to be partitioned into their domains. The most established and robust way to do this, given the current data situation, is by
sequence-based approaches.
The general principle of such methods is based on the assumption that even
if related amino acid sequences diverge during evolution, they normally
maintain the outcome of their important functional parts, the domains.
However, this separation of sequence information complicates the detection of distantly related members. For that reason the established strategy
starts with the maintenance of a diverse set of validated members to catch
the position specific distribution of amino acids. This information is then

brought together in a way that allows for biologically relevant comparison of new proteins, for example using profile hidden markov models (HMMs) [139–141]. On what basis the initial information is chosen and how the data is combined and compared is where all individual methods differ. To annotate the domains within proteins, it appeared worthwhile to combine the strengths of the databases Pfam [142] and InterPro [128].

Pfam

Pfam [109, 142] is one of the oldest and most established approaches in computational biology. Pfam consists of two databases: Pfam-A, where critical steps are manually curated, and Pfam-B, which contains families that are generated automatically using the DOMAINER algorithm [143]. Since Pfam-A domains are of much higher reliability, have permanent accession numbers and are the prevailing standard in the naming of domain types when it comes to domain-domain interactions [111, 112], they are used as the primary identifiers to label domain types throughout the thesis and within the algorithms.

The creation of a Pfam-A domain starts with a handcrafted set of representative seed protein sequences which are used to produce an alignment that is also manually checked. From this initial high-quality multiple sequence alignment, the seed alignment, a profile HMM is generated that is searched against the sequences stored in the UniProt database (UniProtKB) [127], NCBI reference sequences (RefSeq) [144] as well as a set of metagenomic sequences [142]. All sequence regions that score above a family-specific threshold are then included in the final alignment. The thresholds are chosen very conservatively to avoid the introduction of false positives into the alignments.

Pfam annotation of proteins is queried directly from their web servers, which provide the results in the well processable Extensible Markup Language (XML) and, contrary to the Pfam annotation supplied by UniProt, also allows to obtain significance values for each associated domain. The current version Pfam 27.0 (updated March 2013) contains $14,831$ protein families.

InterPro

The InterPro database [110, 128] also aims to divide proteins into their functional modules. Instead of proposing another variation of the already mentioned standard approach to generate domain associations, InterPro aims to combine various complementary existing databases to build its own descriptive signatures. The integrated data sources use amino acid sequence, as well as structural and functional clues to collect representative proteins and employ very different methodologies to infer their own informative signatures. Table 3.1 lists all databases that are currently brought together for InterPro's signature generation process [128].

data source	notes to methodology
Pfam [142]	curated, HMMs, continuous profiles
PRINTS [145]	compendium of fingerprints, not necessarily continuous
PROSITE [146]	regular expressions and profiles, information about structurally or functionally critical amino acids
SMART [147]	curated, HMMs, originally specialized in signaling domains
ProDOM [148]	curated, uses PSI-BLAST [149] and SCOP [150]
PIRSF [151]	curated, whole proteins, evolutionary relationships as networks
SUPERFAMILY [152]	HMMs based on SCOP [150]
PANTHER [153]	curated, profiles/HMMs/family trees, uses pathway and functional annotation
CATH-Gene3D [154]	uses CATH [155]
TIGRFAMs [156]	curated, HMMs, specialized in prokaryotes
HAMAP [157]	curated, profiles, specialized in bacteria

Table 3.1: Databases integrated in InterPro 45.0 (Nov. 2013 release) and their main evidence of initial motif generation [128].

While methods based on the comparison with regular expressions are often unreliable in the identification of highly-divergent super-families, fingerprints (groups of conserved motifs) are a bad choice to detect short functional motifs and profile-based methods, like HMMs, are inferior for sub-family detection [158, 159]. Due to the individual strengths, weaknesses and distinct scopes of the applicable methods, all primary resources have different areas of optimum application and a consolidation of their specific signals is certainly worthwhile.

To build meaningful signatures, all incorporated databases are first put in reasonable parent-child relationships to each other according to their type of detection. For example, PROSITE patterns are typically contained within a PRINTS fingerprint which itself could be enclosed by a Pfam domain. Such hierarchies allow to denote more specific subtypes. For example, different fingerprints or combinations of motifs inside the same detected Pfam domain can help to distinguish sub-sets of more specific variants within sequences that fit a more general pattern [110].

All recent signatures produced by member databases are regularly collected and matched to the current release of UniProt protein sequences using their software InterProScan [159]. If sequence signatures from different sources match the same set of proteins and within each one they overlap in a common region of the sequence, where the previously defined relationships allow, they are assumed to describe the same functional segment and are thus stored together as contributing signatures into a single InterPro entry. This not only allows for consistent naming of equivalent entities across many databases but also increases the overall coverage of the annotated protein space compared to the individual methods [158]. MET4 (UniProt accession P32389), a transcriptional activator of the sulfur metabolism in yeast, for example, is not annotated in Pfam. In InterPro it has one assignment to the entry IPR004827:"basic-leucine zipper domain" that is justified by a SMART match of signature SM00338:"basic-leucine zipper (bZIP)". In both databases these domain entries are described as a basic DNA-binding region followed by a leucine zipper dimerization/protein interacting region. The contributing Pfam signatures of the InterPro entry can be queried as PF00170:"bZIP_1" and PF07716:"bZIP_2", whose functional characterizations match the expectations. As a result of the added sensitivity achieved by the integration of many detection methods, specific domain types can be assigned to the protein instead of being left in the dark.

The InterPro data is retrieved in one go together with other annotation data during the download of UniProt and detected InterPro signatures are converted to their contributing Pfam identifier(s) using the EMBL-EBI 'Dbfetch' web service [160]. The most recent release InterPro 45.0 (updated 19th November 2013) contains 25, 236 entries.

3.2.3.2 Interactions between domains

For comprehensive domain-interaction data the two databases IDDI [111] and DOMINE [112] are combined. Both are integrated databases that additionally apply their own reliability estimations to the data since a large part of the available data consists of rather unreliable predicted interactions.

IDDI [111] amalgamates three structure-based DDI-datasets and 20 computationally predicted ones and complements each interaction with a numerical reliability score that is computed using a weighted overlap methodology by [115]. This method rates each predicted dataset according to a sophisticated score based on the overlap with a gold-standard positive set and combines each evidence. In the original publication a performance evaluation was conducted which suggests a reliability score cutoff to obtain a set of interactions with an accuracy of 90%. The IDDI data of the latest release (May 2011) was downloaded from their website as a simple textfile and restricted to this fraction of most reliable interactions resulting in 23595 DDIs for 5152 domains.

DOMINE [112] integrates two structure-based DDI-datasets and 7 predicted ones whereby all interactions have been classified into categories by a simple classification scheme. Restricted to the interactions inferred from PDB entries and the class of high-confidence predictions, 7588 DDIs for 4344 domains are annotated in version 2.0 (Sept. 2010). DOMINE's entries were also downloaded in a simple textfile from their website.

Although IDDI covers most of of the data supplied by DOMINE the additional integration is still worthwhile. Overall the fused high-confidence dataset consists of 26, 385 DDIs between 6194 domains.

3.2.4 Transcription factors and their binding sites

The Yeast Promotor Atlas (YPA) [161] is a comprehensive database that integrates many data sources of *S.cerevisiae* promotor features. For the thesis only the collection of putative TFs and their binding sites is of further interest. The binding data consists of predictions by MacIsaac et al. [162] (conservation based motif discovery) and several mixed (experimental and predicted evidence) databases: MYBS [163], Swissregulon [164], Yeastract [165] and SCPD [166].

The YPA web interface allows to set a few options to filter binding sites. The data was restricted to binding sites with binding affinity p-values $p < 0.001$ for ChIP-based data, sequence conservation across three species for methods that make use of conservation constraints and predictions involving weight matrices are limited to the most restrictive probability setting $P > 0.5$. The start and end sites for each annotated binding site on a target gene (given as ordered locus names) are gathered per TF (or binding motif, see next subsection) using a custom retrieval tool. This was done for the YPA in version 1.6, 30. June 2012. Subsequent data preparation steps are documented in the upcoming subsections.

Associating binding motifs with proteins

For our analysis it is necessary to link a certain TF with its corresponding target genes and even binding sites. Especially in older literature, binding sites have often been termed by sequence-motifs, not the actual regulatory proteins, since the exact DNA-binding protein was not known at that time. SCPD's data is still organized by motifs and to integrate it the motifs must be mapped to binding protein(s). Most of the time, motifs can be mapped unambiguously to proteins with the previously gathered naming data (for example motif 'HSTF' is the protein 'HSF'), but some need to be curated by hand. If only name conversion was applied, the 'MCB' motif, for example, would be wrongly associated with the protein 'MCB1' ('RPN10_YEAST'), which is a subunit of a proteasome [127], and other motifs cannot even be referred by name. Table 3.2 summarizes which proteins are assigned to motifs and on the basis of which evidence. In two cases, SCB and MCB, the motifs describe binding events of heterodimers. As not to introduce wrong binding site interval annotation for the individual proteins, which would be very problematic when distance constraints are incorporated in the detection of common target genes, those motifs are discarded. Overall 97, 005 interactions between TFs and target genes are included, individual binding sites not counted.

Merging redundant binding sites

When many sources that depict the same kind of data are combined, it often happens that the same event is reported multiple times, i.e. a binding event at basically the same location is reported by at least two different sources. Additionally, since the methods of data acquisition are very heterogeneous

motif	TF(s)	evidence
PAC	DOT6, TOD6	[167]
ATF	ACA1, ACA2	[168]
CRE	SKO1	[169]
MCB	MBP1/SWI6	[170]
SCB	SWI4/SWI6	[171]
CSRE	SIP4	[172]
T4C	not relatable	-

Table 3.2: Table of manually curated associations of sequence motifs with binding proteins and their evidence. If several TFs are referred to a motif they are separated by a comma. If the motif is actually a composite motif of a complex formed by two TFs, they are separated by a slash. Table was also used in supplement of [1].

(various types of experiments/predictions) and binding sites can be defined using different criteria, one has to expect varying length of reported binding events for the same TF. Commonly reported binding site indications are the 'footprint', the patch of DNA protected from nuclease digestion by the TF in the bound state, and the segment of nucleotides that ensures binding specificity. While sites reported by footprints typically span 10-20 bp, the actual specificity is determined by only 5-8 bp of nucleotides [4, 58].

To merge equivalent binding site intervals, an approach similar to sweep-algorithms [97] was used for every TF per gene. All binding sites from all sources are thrown together and stored in a list lt as intervals with a left and right border base pair as given by the data. During the parsing of the raw data the maximal interval size for a TF len_{max} is kept up-to-date. This has linear effort in the number of intervals. Afterwards, lt is sorted by the left interval start in increasing order (log-linear effort). Merging intervals that are sorted has a tremendous advantage: since pairs that are coalescence candidates have to be adjacent, it can be done in linear time by deliberately "sweeping" across lt once. The idea is to iterate using a current interval (l_c, r_c) that is compared to its imminent neighbor with borders (l, r). If the intervals are disjoint the binding sites are treated as independent and the sweep continues with (l,r). If an interval is contained within the other the inner one is discarded. Overlaps are more tricky: if the union of both intervals (l_c,r) spans a binding site within the magnitude of reported intervals (at most len_{max}), so in a sense within the variance of the input data, they are merged. Otherwise they are left separate. Algorithm 3.2 summarizes the procedure.

Algorithm 3.2 Binding site "sweep" algorithm

---

add all BS of an TF for a certain gene as a tuple (l,r) to a list lt thereby keep maximal interval size len_{max} updated {tuple defines interval border}

sort lt by $left$ in increasing order

initialize empty list BS

$(l_c,r_c) \leftarrow$ first element of lt {initialize current running variable}

for (l,r) in lt, beginning with second element **do**

    if $l \geq r_c$ **then**

        append (l_c,r_c) to BS {report independent interval}

        $(l_c,r_c) \leftarrow (l,r)$ {update}

    else if $r \geq r_c$ and $l = l_c$ **then**

        $(l_c,r_c) \leftarrow (l,r)$ {update, since current within later interval}

    else if $r \leq r_c$ and $l \geq l_c$ **then**

        continue {skip since later interval within currrent one}

    else

        if $r - l_c \leq len_{max}$ **then**

            $r_c \leftarrow r$ {expand interval}

        else

            append (l_c,r_c) to BS

            $(l_c,r_c) \leftarrow (l,r)$ {update}

        end if

    end if

end for

append (l_c,r_c) to BS {no comparison or last interval}

return BS {BS contains the cleaned binding site intervals}

Overall the procedure is then upper-bounded by the sorting step and exhibits log-linear effort in the number of intervals (naive: quadratic). Figure 3.5 shows the result of a merging step on the example of MCM1 binding sites on a crowded segment in the promotor of CLN3.

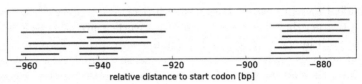

relative distance to start codon [bp]

Figure 3.5: The black line segments denote all the reported binding sites of MCM1 and their locations on the promotor of CLN3. The gray ones are the output of the merging procedure.

After the merging step, $177,135$ interactions between TFs and their exact binding intervals are stored for later usage to estimate colocalization of binding events on target genes.

3.2.5 Expression data

For the subsequent target-based analysis, time-series gene expression profile data is needed. Chin et al. [173] provide a 32 timepoint yeast cell-cycle time-series using up-to-date Affymetrix GeneChip®Yeast Genome 2.0 arrays. Their data was retrieved from the NCBI Gene Expression Omnibus (NCBI GEO) database [174] using accession GSE30053.

Standard microarray preprocessing was conducted in R [175] using the Bioconductor framework [176]. At first the probes referring to *S.pombe* transcripts were masked according to the "s_cerevisiae.msk" mask file from the Affymetrix website. The next step was the normalization. While sophisticated methods like GCRMA [177] are superior when it comes to the accurate measure of differential expression between two or more phenotypes, simpler ones like MAS5 [178] were reported to be better suited for the faithful reconstruction of regulatory networks [179]. The data was preprocessed using the MAS5 implementation in Bioconductor's "affy" package [180]. In the last step the Affymetrix probe ids were mapped to yeast ORFs with the "yeast2ORF" package [181] and the final data table written to a plain textfile for further usage in Python scripts.

An expression-handler class can then be used to load the preprocessed data. The profiles are mapped to UniProt IDs and all appearances of ORFs that are occurring more than once on the array are averaged. By default each individual time-series is centered to its mean and scaled to unit variance to enable the meaningful usage of the euclidean distance in further computations [116, 182].

3.2.6 Reference data

For the evaluation of the approach and the comparison to existing protein complex prediction methods, common complex prediction quality measures are applied (see Section 2.1.2). These necessarily require yeast-specific datasets of known protein complexes, protein localization and associated GO terms.

GO-annotation for each yeast protein can already be provided by the UniProt linkage (Section 3.2.1), the remaining requirements call for special resources.

Yeast reference protein complexes

Classical yeast protein complex reference datasets can be found in the Munich Information Center for Protein Sequences (MIPS) database [130] and the Saccharomyces Genome Database (SGD) [183]. An already preprocessed version of the two databases by Nepusz et al. [81] was downloaded from http://membrane.cs.rhul.ac.uk/static/cl1/cl1_gold_standard.zip as described in [1].

MIPS complexes were derived using the MIPS ComplexCat category during 18th May 2006 with at least 3 and at most 100 members. MIPS category 550 was discarded since it only contains algorithmically inferred complexes. The thus built MIPS reference set contains 204 protein complexes, 26 thereof include TFs annotated in the YPA.

The SGD complex data was derived using gene ontology term annotation based on the entries descending from "GO:0043234" (protein complex) during 11th Aug. 2010. Annotations with negations, colocalization or supported by "IEA" (Inferred from Electronic Annotation) evidence were discarded. Additionally, for the thesis only complexes with at least two members are taken into account, since the prepared source also included complexes of size one. Overall the SGD reference data contains 306 complexes, 17 of them contain TFs.

Since the formerly mentioned databases no longer represent the current state of knowledge in proteomics, additionally CYC2008 [72], an extensive collection of 408 manually curated heteromeric protein complexes from *S.cerevisiae*, is used. This comprehensive and novel catalogue was compiled using literature evidence by more recent small-scale experiments. The reference set was downloaded at http://wodaklab.org/cyc2008/ in version 2.0. Out of 408 complexes, 28 involve TFs.

Modularity of protein complexes is characterized by shared subunits of proteins that participate in more than one functional complex. In annotated complexes such proteins are found in the intersection of two individual sets. We examined the thus prepared gold standards for overlaps within the same dataset. Such overlaps occur for about 44% of CYC2008, 87% of MIPS and 53% of all annotated SGD complexes. Confined to the respective fraction of TF complexes such overlaps are the case for 54% of TF complexes in CYC2008, 88% in MIPS and 56% in the SGD dataset. Combinatorial effects as adressed in the previous Section 2.2 are thus eminently present in the reference data that will be used for evaluation.

Yeast localization data

Huh et al. [184] classified most yeast proteins into 22 distinct subcellular compartments by tagging them with green fluorescent protein and subsequent fluorescence microscopy. The analyzed data is made available at http://yeastgfp.yeastgenome.org/ and covers the associated compartments of 6234 yeast ORFs which can be uniquely mapped to an UniProt identifier using the previously gathered naming data. Among the differentiable localities is also the entity "nucleus". This enables to utilize the mapping to compute colocalization within complexes with relevance to our scope.

3.3   Workflow and implementation

The main algorithm as well as all data retrieval and most data preparation steps were implemented in Python [129]. For the preprocessing of the microarray data, R [175] was used.

The only data sources needed as an input for the algorithm is a probability-weighted protein-protein interaction network in the SIF format with nodes named by their UniProt IDs [126], a list of putative transcription factor proteins that is used to seed the growth, a threshold to generate the seed pairs and an upper bound for the depth of search to keep the combinatorial explosion local. Given those requirements, the tool automatically retrieves all the data that it needs to buildup the domain-domain interaction network

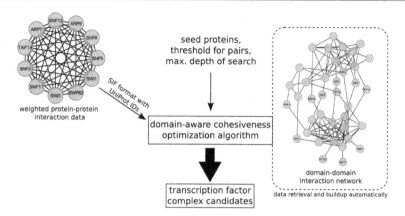

Figure 3.6: The main algorithm together with its necessary input data, computed output and automatically retrieved information is shown. A detailed description also adressing postprocessing possibilities is given in the continuous text. The network examples show the corresponding subnetworks of the SWI/SNF complex (as annotated in CYC2008 [72]) built within the documented framework and visualized using Cytoscape [138]. As can be seen easily, the additional domain network enables a very different view on the connectivity of the complex compared to the perfect clique stretched by high probability edges in the PPIN. Figure adapted from [1].

using plenty of diverse data sources and runs the domain-aware cohesiveness optimization algorithm to finally predict TF complex candidates. This part of the thesis is completely independent of the organism and can directly be applied to any other desired organism if the required input data is available.

Figure 3.6 gives an overview over the general workflow. Given a PPIN, a corresponding DDIN is built automatically, using the data sources discussed before, and the algorithm is executed for all determined pairs. A weighted and an unweighted version of the algorithm are proposed, while the unweighted version allows for a more radical runtime optimization. Entries within the final output that are subsets of other entries are discarded since they are not optimally enlarged. Furthermore, one of several merging approaches can optionally be added as an additional postprocessing step. The exact implementations of all those individual parts will now be discussed in detail. The last paragraph additionally covers the implementations of batch GO annotation enrichment analysis and expression coherence scoring that are needed later during the evaluation.

3.3.1 Building the domain-domain interaction network

The automatic construction of a DDIN from a given PPIN using our data sources is straightforward. For every protein in the PrePPI-PPIN Pfam is queried for the primary domain assignment since Pfam also reports multiplicity of domains. An InterPro request is submitted additionally and its matches are translated to their contributing Pfam identifiers. Domains that have not been detected using Pfam are supplemented by the extra query. The integration of the second source led to a better annotation for about 11% of the proteins that span the PrePPI-network (704 of 6698 proteins, roughly in the range of structurally known yeast proteins).

The next step is to establish all applicable interactions between domains. Using the combined IDDI/DOMINE dataset, edges between all interacting domains among different proteins in the network are established if the protein pair is also connected in the PPIN. This ensures consistently having a corresponding PPI to every DDI which can, for instance, later be used to assign probabilities to domain interactions.

The six putative uncharacterized proteins YA37B, YAJ9, YO011, YB223, YBO2 and YO134 were filtered out at this stage because their only reported protein interactions were self-interactions.

As motivated by Ma et al. [94] (see Section 2.2.1), artificial domains are introduced per default to maintain the connectivity of the network. Here, an extra domain is initially added to every protein and utilized to establish a link between interaction partners that are listed in the PPIN but have no associated edge in the DDIN after incorporation of the domain interaction step. This helps to further ensure the consistency between DDIN and PPIN despite the missing data but, unlike their original approach, in a deterministic way.

The domain-interaction network obtained by the automatic construction approach applied to the PrePPI protein-interaction network contains 6692 proteins, $11,875$ domains (on average about two domains per protein) and $314,182$ domain-interactions.

3.3.2 Domain-aware cohesiveness optimization

The core principle was already outlined in Section 3.1. To tame the run-time the implementation is tuned by considering several problem-adjusted algorithmic details.

The actual growth procedure is implemented in a depth-first manner and memorizes processed subtrees by storing a hash value based on a state definition to ultimately avoid revisiting the same state several times. A modified state description will be proposed that potentially pronounces the positive effect of such an optimization by memoization.

In the non-terminating steps of the depth-first-like progression, tedious constant computational costs can be saved by consequent reusage of already determined parts of the cohesiveness calculation throughout every subtree and a justified filter discards unnecessary branches during expansion steps. Finally, a variant of the main algorithm will be introduced that is able to utilize the weight annotation of the PrePPI data as an early pruning criterion.

Constant costs during each step

w_p^{in} denotes the total weight of edges that connect protein p with members of V and $w^{in}(V)$ is the total weight within members of V. Additionally, we define w_p^{out} as the total weight of edges that connect protein p with non-members of V and $w^{bound}(V)$ as the total boundary weight of members in V.

As is shown in the supplement of reference [81] and [1], the cohesiveness can be rewritten to account for the change inflicted by the addition of a single protein:

$$w^{in}(V \cup \{p\}) = w^{in}(V) + w_p^{in}$$

$$w^{bound}(V \cup \{p\}) = w^{bound}(V) - w_p^{in} + w_p^{out}$$

$$f(V \cup \{p\}) = \frac{w^{in}(V \cup \{p\})}{w^{in}(V \cup \{p\}) + w^{bound}(V \cup \{p\})}$$

$$=^{use} \frac{w^{in}(V) + w_p^{in}}{w^{in}(V) + w^{bound}(V) + w_p^{out}}$$

And the same can be done for the removal of a protein [81]:

$$w^{in}(V \setminus \{p\}) = w^{in}(V) - w_p^{in}$$

$$w^{bound}(V \setminus \{p\}) = w^{bound}(V) + w_p^{in} - w_p^{out}$$

$$f(V \setminus \{p\}) = \frac{w^{in}(V \setminus \{p\})}{w^{in}(V \setminus \{p\}) + w^{bound}(V \setminus \{p\})}$$

$$= \frac{w^{in}(V) - w_p^{in}}{w^{in}(V) + w^{bound}(V) - w_p^{out}}$$

The equations show that the cohesiveness does not necessarily need to be computed from scratch in every single step of the algorithm, it suffices to incorporate the impact of the induced change. The implementation uses this property and thus separately stores the current internal weight $w^{in}(V)$ and the boundary/outer weight $w^{bound}(V)$ to enable the usage of the above formulations. This modification decreases the computational effort to compute the cohesiveness for every change by one protein from quadratic to linear in the size of V [1].

Algorithm 3.3 illustrates the extended step function considering this feature. The algorithm can also be found in the supplement of [1].

Filter unnecessary branching

The computationally most expensive steps are the branching ones with many realization possibilities, so one should avoid to take unnecessary branches into account.

Many proteins have the same domain type repeatedly, so-called domain multiplicity. Such proteins evoke an overestimation of necessary domain choices: we only need to compute one of these choices, not all variants, because all connections of these domains can also be realized by all other domains of the same type within the protein. We just need to know the consequences of blocking one of them since they have the same interactions to the outside. Figure 3.7 shows an example for such a case.

The filter is applied by default when DDI choices are determined.

Algorithm 3.3 Faster domain-aware cohesiveness optimization step function:
step(V, E, cw_{in}, cw_{out}) (first part)

given current proteins V, used domain interactions E and current inner/ outer weights of V {as described later, actual state description can differ}

determine V_{inc} and V_{bound} from the DDIN and the used edges E
$csum \leftarrow cw_{in} + cw_{out}$
$max \leftarrow cw_{in}/csum$
$(nw_{in}, nw_{out}) \leftarrow (0.0, 0.0)$ {the nw_i account for the Δ induced by best move so far}
$action \leftarrow$ terminate

for $\forall P \in V_{inc}$ **do**
  $(w_{in}, w_{out}) \leftarrow$ computeDelta(P, V) {get induced change in inner/boundary weights}
  $coh \leftarrow (cw_{in} + w_{in})/(csum + w_{out})$
  if $coh > max$ **then**
    $max \leftarrow coh$ {update max. references}
    $action \leftarrow$ add P
    $(nw_{in}, nw_{out}) \leftarrow (w_{in}, w_{out})$
  end if
end for
for $\forall P \in V_{bound}$ **do**
  $(w_{in}, w_{out}) \leftarrow$ computeDelta(P, V)
  $coh \leftarrow (cw_{in} - w_{in})/(csum - w_{out})$
  if $f(V') > max$ **then**
    $max \leftarrow f(V')$
    $action \leftarrow$ remove P
    $(nw_{in}, nw_{out}) \leftarrow (w_{in}, w_{out})$
  end if
end for

(continued in 3.4)

Algorithm 3.4 Faster domain-aware cohesiveness optimization step function:
step(V, E, cw_{in}, cw_{out}) (second part)

---

if *action* = terminate **then**
    return result V {termination yields a complex candidate}
else if *action* = remove P **then**
    $V' \leftarrow V \setminus \{P\}$
    determine $e \in E$ that connects P to V' in DDIN
    return job $(V', E \setminus \{e\}, cw_{in} - nw_{in}, cw_{out} + nw_{in} - nw_{out})$ {removal yields the state description for the next step, updated inner/boundary weights to match V'}
else if *action* = add P **then**
    $V' \leftarrow V \cup \{P\}$
    initialize empty list l
    $(tw_{in}, tw_{out}) \leftarrow (cw_{in} + nw_{in}, cw_{out} - nw_{in} + nw_{out})$ {update the change in weights for V'}
    for $\forall e \in$ DDIN so that $E \cup e$ spans V' **do**
        append $(V', E \cup e, tw_{in}, tw_{out})$ to l
    end for
    return list of jobs l {addition yields a list of state descriptions that need to be further processed}
end if

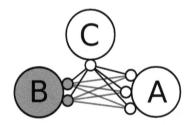

Figure 3.7: In this example domain network all domains of proteins A and B are of the same type and both types are able to interact with the domain of protein C. Currently $V = \{A\}$ and B should be added. At first sight, one has six possibilities to connect B to A in this case (edges marked gray). In general, the choice of the edge will matter in later iterations, in the case of domain multiplicity using any of these edges is sufficient to cover all cases. Thus such a filter discards five branches in this example.

3.3.2.1 Avoid revisiting states and optimized state description

In a graph-based growing algorithm it is very likely that we repeatedly grow into the same local maxima that we already hit during the processing of previous startpoints and struggle with unnecessary duplicated computations. For example, one could obtain an intermediate result collecting proteins {ABC} coming from {AB} and also from {BC}, leading to identical progression from this early point on for different branches in the course of a run of the algorithm. If we encounter an identical algorithmic state a second time, we should not proceed further since we do not want to solve the same subproblem twice. The exact definition of a state will also be of importance for the problem.

Using an efficient set data structure based on hash functions one can memorize and lookup states that have been processed before mostly in constant time, an optimization termed memoization [97]. Hash function-based sets are the default implementation in CPython (the 'standard Python') and equivalent to HashSets in Java. The iterative depth-first manner of the overall approach makes the administration of such a state memoization rather easy. An alternative strategy would be, for example, to first grow triples of proteins from all pairs, compare them among each other for duplicates and move on accordingly.

Algorithm 3.5 describes the implementation in a generalized way independent of the actual state description.

As a remark, in principle it would be easy to program a concurrent version of this main loop, in practice a parallelized Python implementation is often problematic. Due to the "global interpreter lock" in Python, a mechanism that enforces single-threaded computation on the lowest level of the interpreter, concurrency that utilizes all physical cores can only be achieved by simultaneous execution of several instances of the interpreter [185]. Since the memory resources per instance are exclusive (distributed memory), tedious exchange and synchronization of all cached data using interprocess communication would be necessary. The domain-aware cohesiveness optimization algorithm is not implemented in parallel to better account for the actual impact of the optimization by memoization (no communication overhead); several other parts of the thesis are.

The smallest obvious state definition is given by the set of interactions that are chosen on the domain-level in each step. These interactions contain the

Algorithm 3.5 GrowthManager(*start*, *cut*)

DDIN object has attribute *hash_memo* {which is used to memoize visited states across all startpoints}
initialize empty set *results*
initialize stack *stack* with starting pair state *start*
while |*stack*| > 0 **do**
    V, *state* ← *stack*.pop() {process job}
    str_hash ← hash of state *state*
    if *str_hash* ∈ *hash_memo* **then**
        continue {state already processed, discard}
    else if |*V*| = *cut* **then**
        results.add(*V*) {if size cutoff is attained, terminate}
    else
        result ← step(*V*, *state*) {process job, the iterative part on a generalized step function allowing different state descriptions}
        hash_memo.add(*str_hash*) {mark this subtree as processed}
        if *result* is a set **then**
            results.add(*result*) {job terminated and returned actual complex candidate}
        else if *result* is a list **then**
            push content of *result* onto *stack* {job needs to branch, returns all further state descriptions}
        else
            push *result* onto *stack* {job has only one branch, returns this state description}
        end if
    end if
end while
return *results*

information of proteins in the current dense set, their exact domain connectivity and occupancy. Such a definition contains all information that is needed to implement addition and removal of proteins with ease.

However, since memoization is always a space-time tradeoff, one should try to get most out of its utilization. To make the effect of falling into previously seen states more likely one can further adjust one's definition of a state in a way that is adjusted to our specific problem instance. In this greedy growing procedure the very most of steps will suggest the addition of a protein. To add a protein the optimization algorithm does not need to know the exact domain-level topology of the current proteins, it is only interested in the resulting domain occupancy. Both descriptions, exact edges as well as occupied domains, lead to the same solutions because expansion only cares about domains that are still usable. However, occupied domains, in the sense of a state definition, are a superset of a state definition involving the chosen edges on the DDIN since many different edge choices could block the same domains. Thus the chance to hit a state repeatedly becomes higher since multiple states are condensed to one while the basic invariant of a consistent spanning-tree is preserved in this 'honeypot'-definition.

Of course the algorithm still needs to behave correctly for all possible outcomes of each step, not only the addition of proteins. Since we are losing the exact connectivity-information on our path while using the optimized state definition, we can run into problems when cohesiveness suggests to remove a protein P. As we still know which domains are occupied and that P needs to be connected by a domain interaction to the other proteins in V, it is possible that only one edge e in the DDIN connects the single occupied domain of P with the occupied domain of another protein in V. Then we can use this distinct interaction e to free the associated domains, remove P from V, and proceed correctly.

If there is no distinct interaction e we run into an ambiguous choice of several domain interactions that could have potentially caused a certain domain occupancy. Under these circumstances the invariant of a spanning-tree in the DDIN cannot be preserved reliably and we have to fall back to the untuned state definition of the algorithm and rerun the calculations for the startpoint. Although we have to redo the whole calculation the effort stays the same asymptotically.

Figure 3.8 demonstrates possible cases on an example.

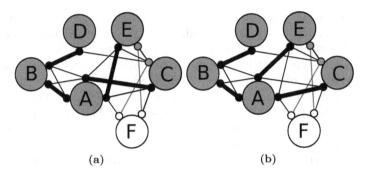

(a)   (b)

Figure 3.8: This example shows two distinct execution conditions of the algorithm for the same cluster in the DDIN. Despite differing active (chosen) domain interactions (thick edges) that connect the internal proteins in V, the two conditions show the same occupancy of domains (marked in black) and thus in turn the same domains of internal proteins are available (marked in gray). So, depending on the state definition used for memoization, (a) and (b) are assumed to be equal or they are not. Notwithstanding the past history of the two conditions, they both allow for exactly the same choices when it comes to the addition of a protein. If, however, a protein should be removed from V, the past history can matter. If B should be removed from V this can be done unequivocally in both state definitions because only one interaction of its occupied domain leads to another occupied domain. If D or F should be removed the sole domain occupany only provides ambiguous information for this topology. For both proteins it cannot be inferred which of C's domains needs to be freed if the exact interactions are concealed.

The implementation so far only needs slight adjustments to account for this state condensation feature: Algorithm 3.5, also found in the supplement of [1], already outlines the essential proceeding. A state condensation variant of the algorithm will first try to optimize a starting pair using the "occupied domains" state description. If, however, an ambiguous removal occurs, the algorithm needs to switch back to the slower "domain interactions used" state description to have enough information available. The benefit of this optimized state description on actual biological data will be evaluated later.

3.3.2.2 Using weights to bound branching

Another effective consideration is to exploit the weight annotation of PrePPI to associate each DDI with the probability of the corresponding edge between the connected proteins in the PPIN to establish an overall probability that can be used for an early pruning criterion.

While this correspondence is potentially error-prone since the probability will likely represent an upper limit achievable by any of the domain inter-actions between the proteins (or even a combination of several) but not the clear contribution of the individual choice, such an overestimation of the actual probability would even provide a conservative estimate when utilized for comparison with a lower bound threshold.

Under the assumption that spanning edges in the DDIN are independent of each other, the probability to observe an underlying spanning tree on the domain level is the product of the probabilities of all its spanning edges. If, during the execution of the step function of the algorithm, a resulting cohesiveness-maximizing candidate would yield a value below 0.5, the current cluster is returned instead. In the case of an addition, all alternatives below the threshold are withdrawn and the current cluster is only returned if no alternative choice shows a total probability within the boundary.

Because every current complex candidate is assigned a probability value inherently based on its exact spanning edges and not the domain occupancy pattern, the state-condensation trick as suggested earlier can not be applied here and one needs to be content with the edge-choice memoization. Also, the results will, of course, differ from those of the implementations seen so far since the early pruning will not allow unlikely candidate complexes to grow any further.

Algorithm 3.6 introduces the idea into the already shown function. It can also be found in the supplement of [1].

Merging of highly overlapping complex candidates

To estimate the additional value of the possibility to predict larger complexes and to unify highly overlapping complexes, three different merging approaches are additionally implemented to be used as a potential post-processing step after the actual main algorithm: a simple method based on the exact postprocessing as used by [81] and two versions of an iterative approach that additionally imposes the DDI-model as a filter in the fashion of [94] (explained in Section 2.2.1).

The simple method constructs an overlap graph - a graph where predicted complexes are the nodes and are connected by an edge if the complex pairs fulfil a certain overlap criterion - and amalgamates all complexes within connected components of this graph to their union of proteins. Overlaps are defined to have an overlap score ω above 0.8.

Algorithm 3.6 Domain-aware cohesiveness optimization step function, with weights:
wstep(V, E, cw_{in}, cw_{out}, cP)

---

as before but current probability cP additionally given

... as before ...

if $action$ = terminate **then**
   return result V
else if $action$ = remove P **then**
   $V' \leftarrow V \setminus \{P\}$
   determine $e \in E$ that connects P to V' in DDIN
   $nP \leftarrow cP/$ (weight of e in PPIN)
   return job $(V', E \setminus \{e\}, cw_{in} - nw_{in}, cw_{out} + nw_{in} - nw_{out}, nP)$
else if $action$ = add P **then**
   $V' \leftarrow V \cup \{P\}$
   initialize empty list l
   $(tw_{in}, tw_{out}) \leftarrow (cw_{in} + nw_{in}, cw_{out} - nw_{in} + nw_{out})$
   for $\forall e \in$ DDIN so that $E \cup e$ spans V' **do**
      $nP \leftarrow P*$ (weight of e in PPIN)
      if $nP \geq 0.5$ **then**
         append $(V', E \cup e, tw_{in}, tw_{out}, nP)$ to l {add only if above probability threshold}
      end if
   end for
   if $|l| = 0$ **then**
      return result V {if no branch above threshold, terminate}
   end if
   return list of jobs l
end if

The two other algorithms additionally incorporate the DDI-model filter as suggested by [94] and iteratively unite highly overlapping candidates pairwise to account for finer details. This is done until no further condensation of predicted complexes is applicable. Both variants only differ in the exact matching problem they solve and are therefore explained in one go. While the first one solves the maximum matching problem as done in the original publication, the second one is implemented to use the maximum weight matching to potentially profit from the weight annotation in the PrePPI-PPIN that is again mapped to domain interactions. Algorithm 3.7 explains the iterative procedure whereas Algorithm 3.8 covers the pairwise merge function. The batch of pairwise calculations is processed in parallel during every pass to obtain optimal utilization of the computational resources and the Python package NetworkX [186] is used for efficient implementations of the matching algorithms based on Edmond's matching algorithm [114].

Algorithm 3.7 Iterative merging procedure

$G \leftarrow$ buildOverlapGraph(complex candidates) {using overlap score > 0.8 criterion}
$N \leftarrow$ set of nodes in G
$E \leftarrow$ edges in G
while $|E| > 0$ **do**
  $merged \leftarrow$ all non-empty protein sets from PairMerge(A,B) applied to $\forall\ (A,\ B) \in E$
  $G \leftarrow$ buildOverlapGraph($merged$) {again search for overlaps within the merged clusters}
  add nodes in G to N {bookmark every candidate that appeared}
  $E \leftarrow$ edges in G {update list of conceivable merging candidates}
end while
discard entries in N that are subsets of other entries in N
return N

Algorithm 3.8 PairMerge(A,B)

$m \leftarrow A \cup B$
if cohesiveness of m is below the individual cohesiveness of A or B **then**
    return empty set {merged pair not fertile}
end if
build DDIN ddi for proteins in m
$match \leftarrow$ compute appropriate matching in ddi {depending on method:
max. match./max. weight. match.}
$cc \leftarrow$ connected components of $match$ in ddi
if $|cc| > 1$ or $|cc| = 1$ but component not equal to m **then**
    return empty set {merged pair not valid in DDI-model}
end if
return m {union m shows increased cohesiveness and is valid in the
DDI-model}

Identifying targets and binding site relations

Binding site associations within the program are internally stored in a
nested map $target\_map$ where each TF i is related to a second map
$target\_map[i]$ associating target genes with their corresponding sorted lists
of binding intervals.

Thus $target\_map$[TF1][GENE1] returns a list of binding event intervals of
TF1 binding to GENE1. Common target genes of a set of TFs S without
any site restrictions can then be easily computed as the intersection of the
keys in the nested second map

$$\text{common targets}(S) = \bigcup_{\forall i \in S} target\_map[i].\text{keys}() \ .$$

In the context of TF cooperativity by complex assembly, good estimates
exist for upper bounds of pairwise distances between utilized binding sites
(see Section 1.2.2). These can be used to further filter the set of potentially
targeted genes by additional integration of the exact binding site annotation. To find segments in which a set of TFs allows pairwise distances
between their given binding sites that are strictly within a certain range of
minimal and maximum base pairs requires a little more effort.

The general procedure for arbitrary sets of TFs relies on the special case
of two TFs. For all genes that are common targets of both factors their

binding sites are compared pairwise and lists of constraint-compatible pairs per shared target gene are returned as shown in Algorithm 3.9.

Algorithm 3.9 Pairwise compatibility:
PWcomp(TF1, TF2, d_{min}, d_{max})

---

given TFs TF1 and TF2 and distance constraints d_{min} and d_{max}
initialize empty map *out*

for $\forall gene \in$ common targets({TF1, TF2}) **do**
  list1 $\leftarrow$ *target_map*[TF1][*gene*] {get binding intervals of TF1 and TF2 on *gene*}
  list2 $\leftarrow$ *target_map*[TF2][*gene*]
  for $\forall$ ($l1$, $r1$) $\in$ list1 **do**
    for $\forall$ ($l2$, $r2$) $\in$ list2 **do**
      initialize empty list *intervals*
      if $l1 \geq l2$ **then**
        if $l1 < (r2 + d_{min})$ or $l1 > (r2 + d_{max})$ **then**
          continue with next interval {invalid overlap or too distant}
        end if
        append $[(l2, r2), (l1, r1)]$ to *intervals* {report matching pair, TF1 right of TF1}
      else
        if $l2 < (r1 + d_{min})$ or $l2 > (r1 + d_{max})$ **then**
          continue with next interval {invalid overlap or too distant}
        end if
        append $[(l1, r1), (l2, r2)]$ to *intervals* {report matching pair, TF1 left of TF1}
      end if
    end for
  end for
  if $|intervals| > 0$ **then**
    out[*gene*] $\leftarrow$ *intervals*
  end if
end for
return *out* {*out* contains the compatible binding site interval choices given the constraints}

Using compatible pairs of binding events as the base case, such a preselection can be extended to include triplets and larger tuples by an additional function that matches possible binding intervals of a new TF into a predetermined list of compatible interval choices per target gene if possible. Nonvalid intervals are removed at an early stage as shown in Algorithm 3.10.

Combining those two basic functions enables to implement the most general case easily. Starting from all distinct pairs of TFS and for all permutations of remaining TFs one by one, all binding site combinations that are valid according to the distance constraints are determined as in Algorithm 3.11. Again, invalid branches are pruned as early as possible.

Since all permutations are tried and all valid combinations are reported, the result is not fixed to some best arrangement but consequently captures every possible solution across all targets. This allows for flexible layouts by the same set of TFs as suggested in recent models (see Section 1.2.2). To allow for slight overlaps or even check for colocalization in this scheme, the minimal distance constraint can simply be set to a negative value.

Expression coherence score and its change

The expression coherence score (ECS) enables to potentially integrate various types of expression data - regardless of whether time-series or steady-state, condition or cell-type specific - into a (conditional) measure of coexpression for sets of genes in a certain cellular state [116]. Additionally, the significance of the score itself, but also its change when assuming interplay between several TFs, can be calculated [116, 123].

The ECS calculations follow the essential methodology as suggested by Pilpel et al. [116]. Contrary to their usage, the correlation is used as a distance measure, binding site constraints can be incorporated in the target gene selection and all computations are implemented to work for arbitrary sized tuples of TFs.

As described in Section 2.3.1, the threshold D is determined as the 5th percentile of pairwise distances across 100 randomly sampled genes from the complete expression data. By default the average of $10,000$ independent runs is taken to obtain a converged threshold D. The ECS for arbitrary sets of genes can then be calculated as the fraction of pairwise distances below D. The significance of an ECS is given as the fraction of randomly drawn gene sets of equal size with at least the same ECS as described earlier

Algorithm 3.10 add to previous candidates:
addTF(TF, *previous, gene, d_{min}, d_{max}*)

given a TF, fixed possibilities for extension *previous* for a given target
gene and the constraints
list2 $\leftarrow$ *target_map*[TF][*gene*] {get binding intervals of TF on *gene*}
initialize empty list *intervals*
for $\forall$ fixed_positions $\in$ *previous* **do**
  for $\forall$ ($l2$, $r2$) $\in$ list2 **do**
    valid $\leftarrow$ False
    for $\forall$ ($l1$, $r1$) $\in$ fixed_positions **do**
      if $l1 \geq l2$ **then**
        if $l1 < (r2 + d_{min})$ **then**
          valid $\leftarrow$ False {overlap, not usable for extension within this
          given interval layout}
          break
        else if $l1 > (r2 + d_{max})$ **then**
          continue {too distant, but probably adjacent (enough) to the
          next fixed binding interval}
        end if
        valid $\leftarrow$ True {just fine}
      else
        if $l2 < (r1 + d_{min})$ **then**
          valid $\leftarrow$ False
          break
        else if $l2 > (r1 + d_{max})$ **then**
          continue
        end if
        valid $\leftarrow$ True
      end if
    end for
    if *valid* **then**
      new $\leftarrow$ ($l2$, $r2$) inserted into proper position within previously
      determined fixed_positions
      append *new* to *intervals* {store the extended valid binding inter-
      val including TF}
    end if
  end for
end for
return *intervals*

Algorithm 3.11 Determine distance constraint adjacent binding sites for a set of TFs:
getBSPossibilities(TFs, d_{min}, d_{max})

---

initialize empty map out
initialize empty map pw
$common \leftarrow$ common targets(TFs)
for $\forall$ distinct pairs of (TF1, TF2) $\in$ TFS **do**
    $pw[(TF1, TF2)] \leftarrow$ PWcomp(TF1, TF2, d_{min}, d_{max}) {precompute pairs for all their targets}
end for
for $\forall target \in common$ **do**
    initialize empty list $intervals$
    for $\forall$ distinct pairs of (TF1, TF2) $\in$ TFS **do**
        $start \leftarrow pw[(TF1, TF2)]$
        if $target \notin start$ **then**
            continue {if current pair has no valid intervals no need to look further}
        end if
        for $\forall$ permutations $perm$ of TFs\{TF1, TF2} **do**
            $res \leftarrow start[]target]$
            for $\forall$ TF $\in perm$ **do**
                $res \leftarrow$ addTF(TF, res, $target$, d_{min}, d_{max})
                if $|res| = 0$ **then**
                    break {no valid choices after trying to extend, no reason to look further}
                end if
            end for
            if $|res| > 0$ **then**
                append all valid solutions in res that are not yet in $intervals$
            end if
        end for
    end for
    if $|intervals| > 0$ **then**
        $out[target] \leftarrow intervals$ {if there are feasible arrangements, store them}
    end if
end for
return out {out contains all compatible binding site interval choices per target gene}

and suggested by [123]. Pairwise distances of expression profiles among set members are computed efficiently using the 'pdist' function of the popular SciPy (Scientific Python) library [187]. There, the correlation as a distance measure is defined as $d_{corr}(A, B) = 1 - corr(A, B)$. The Monte Carlo sampling experiments are consistently implemented to run massively parallel.

The implementation of the significance associated with the change in coherence dECS, induced by the additional restriction of common target genes by a set of TFs compared to the sets that contain one TF less than this, is done in a more generalized way than described in Section 2.3.1 and by [116] and is therefore explained in more detail here. The original approach computes the significance of dECS for pairs of TFs by comparing them to the ECSs of their individual target genes, where target genes are defined by known binding sites on the promotor and mutual targets of two TFs are defined as the intersection of their targets, disregarding any conceivable biological constraints by distance relations of corresponding binding sites [116]. The compiled data and the aid of previously described methods provide the basis for these calculations. Furthermore, the implementation should be able to handle arbitrary sized n-tuples of TFs.

Initially the ECS maximizing TF subset sub_{max} of size $n - 1$ (one TF removed) is determined among all those subsets which differ from the original set of TFs by one TF t. The addition of t to sub_{max} induces an additional constraint on the target genes, changing the set of genes evaluated in the calculation of the ECS of all common targets or, more general, all common target genes where binding site relations are fulfilled. Figure 3.9 visualizes the check for the case of three TFs and explains why a slight redefinition of dECS is necessary to capture the most general case in a correct manner. The random sampling is then carried out using the common targets of sub_{max} according to common binding or constraints.

Algorithm 3.12 shows the procedure for a variant covering binding site constraints. A variant only taking common target genes into account can be obtained by simply replacing getBSPossibilities() by the function calculating common targets.

Algorithm 3.12 Compute significance of dECS with binding site constraints:
dECS(TFs, T, d_{min}, d_{max})

---

given set of TFs, number of random samples T and constraints
$all \leftarrow$ getBSPossibilities(TFs, d_{min}, d_{max}).keys()
if $|all| < 10$ **then**
   return 1.0 {as in original method: discard computations with less than 10 target genes}
end if
$EC_{max} \leftarrow 0$
$sub_{max} \leftarrow$ None
for $\forall t \in$ TFs **do**
   $sub \leftarrow$ TFs\{t} {determine best subset with size $|$TFs$| - 1$}
   $S \leftarrow$ getBSPossibilities(sub, d_{min}, d_{max}).keys()\all
   $EC \leftarrow$ ECS(S)
   if $EC > EC_{max}$ **then**
      $EC_{max} \leftarrow EC$ {update values accordingly}
      $sub_{max} \leftarrow sub$
   end if
end for
dECS $\leftarrow$ ECS(all)$-EC_{max}$ {reference gain in coherence}
$A \leftarrow$ list of getBSPossibilities(sub_{max}, d_{min}, d_{max}).keys() {unrestricted list of common targets for sampling}
initialize empty list $distr$
for T times **do**
   shuffle A {MC sampling is actually distributed to arbitrary number of cores}
   $d_1 \leftarrow$ first $|all|$ entries of A
   $d_2 \leftarrow$ entries after the first $|all|$ of A
   append $d_1 - d_2$ to $distr$ {generate distribution of random splits}
end for
$dT \leftarrow$ number of random samples in $distr \geq$ dECS
return $\max(\frac{1}{T}, \frac{dT}{T})$ {return significance (lower bounded)}

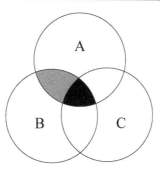

Figure 3.9: Venn diagram visualizing the overlapping target gene sets of TFs A, B and C. Assuming $A \cap B$ (union of colored areas) is the subset of size two with the largest ECS (sub_{max}), then the change of the coherence score dECS is per definition the difference between ECS($A \cap B \cap C$) (common targets of all, black) and ECS(($A \cap B) \setminus C$) (difference induced by addition/removal of constraint C, gray). Since C's targets can only overlap with $A \cap B$ if they are in $A \cap B \cap C$ dECS can be rewritten to dECS = ECS($A \cap B \cap C$) - ECS(($A \cap B) \setminus (A \cap B \cap C)$). This formulation is favourable in practice when it comes to a target gene definition utilizing pairwise distance constraints, since only the notion of compatible targets of all TFs together can reflect the actual set needed for the calculation. Just discarding targets of C does not meet the requirements of the imposed distance constraints since no sufficient information of binding event adjacency is incorporated.

Recapitulating, dECS is the difference in coherence between the green and blue sets (as defined earlier). Its significance is the probability to achieve an equally good or better change with random partitioning of the union of both sets.

Remaining parts

Batch gene enrichment analysis is carried out with a self-made wrapper for the binaries of the open-source software goatools [188] using the GO structure provided by the full OBO ontology file from Dec. 13., 2013, the terms as annotated in UniProt and the whole organism as the background. Significance levels are adjusted using Bonferroni correction. Plots of any kind are generated using the Matplotlib library [189].

Results and Discussion

The first subject of the chapter will be an evaluation of the performance gain that is achieved by the proposed algorithmic optimizations in practice. The remaining chapter will then deal with the actual output. The evaluation and analysis of the predicted TF complexes is split into two major parts:

1) The assessment of the predicted complexes on the basis of common protein complex prediction benchmarks in comparison with popular generic approaches and conceivable merging procedures as a postprocessing step.

2) Based on the results of the previous part, a set of predicted TF complexes will be selected for further analysis. This will include an assignment of possible target genes based on the binding site association of involved TFs and inference of the regulatory influence on the basis of recruited proteins. Additionally, obtained TF tuples are submitted to thorough expression coherence investigation to see if reasonable regulatory effects can be perceived in cell cycle expression data.

All results of the algorithm described in the thesis were obtained as illustrated in Figure 3.6. Given the PrePPI yeast network and the 148 TFs as annotated in the YPA, the complete DDIN was built and seed pairs with a probability of at least 0.75 (or at least two partners if no interaction was within the threshold) were generated. Two TFs annotated in the YPA, MATA1 and MAL63, are not represented in the PrePPI-PPIN and were therefore omitted. Both are not included in the sequence of the reference yeast strain S288C [183, 190]. The thus generated start seed comprised 1526 distinct protein pairs which form 1898 start states considering the choices of the necessary domain interaction between each pair. The productive runs were then executed with a depth threshold of 10, i.e. the proposed algorithm considered complexes containing up to 10 proteins. Lower thresholds were only used to assess the influence of different optimization approaches.

4.1 Impact of algorithm engineering on runtime

To measure the impact of the more sophisticated optimization efforts in practice, three different variants of the proposed TF complex prediction approach were compared. They all possess the same tweaks as illustrated in Section 3.3.2 but differ in their handling of states and state memoization. The second version additionally imposes a probability threshold on the underlying spanning tree in the domain network and the last one uses the proposed state condensation. All measurements were based on the average of three independent runs of the algorithms.

Figure 4.1 shows the runtimes of all executions for different depth thresholds across all variants. Other user-defined parameters were held fixed as described in the beginning. As can be clearly seen, the state condensation

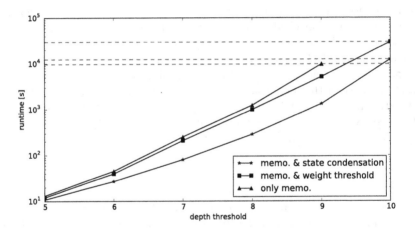

Figure 4.1: The runtimes of three different implementation variants were compared for up to six different depth thresholds. The results are visualized in a semi-log plot, since the combinatorial algorithm is inherently exponential. For practical reasons, the least optimized version was not evaluated with a threshold of 10.

as suggested in Section 3.3.2.1 is not only beneficial in theory, but also on real biological data (at least in this case study). It outperformed the other variants across all the evaluated problem sizes. The difference to the non-tuned state description for the depth threshold of 9 was almost a whole magnitude (9710s vs 1326s). The approach with the early pruning consistently required runtimes in between the two other approaches (5165s). As a remark, only 146 ambiguous remove events - states that force the state condensation approach to restart a certain tree (see Section 3.3.2.1) - occurred with the largest investigated problem size up to depth 10.

An investigation of the slopes, the exponents displayed by the algorithms, can reveal a finer-grained assessment of the situation. The slope of the slowest variant was steadily increasing with increasing problem size, clearly visible as a kink in the semi-log plot. This was also the case when state condensation was applied, but then base steepness and the increase are smaller. With the probability-cutoff, however, the trend seemed to behave slightly different. While the line seemed to proceed almost parallel to the slowest variant for half of the plot, the exponent obviously did not increase any further. Comparing the average runtimes of the two faster candidates at a depth threshold of 10: $12,413$s and $29,730$s (factor about 2.4), with the runtimes at threshold 9: 1326s and 5165s (factor about 3.9), confirmed the visual impression with numerical evidence. This suggests that the addition of a weight-based threshold most likely outperforms the tuned state descriptions in terms of runtime for larger maximal depths. This seems reasonable insofar, as the total probability of a complex candidate during the growth can only decrease with its size which makes pruning events more likely.

State condensation as well as the early pruning criterion based on the probability annotation of the PrePPI data both rendered to be worthwhile sophisticated extensions to the algorithm in terms of runtime. Since their calculations yield different results, their predictive performance should still be examined independently.

4.2 Common protein complex prediction benchmarks

The weighted (referring to the use of a PPIN-weight utilizing internal threshold as mentioned before in Section 3.3.2.2) and unweighted version of the approach that emerged from this thesis were compared to the most recent versions of three popular complex prediction tools: MCODE [80], MCL [89] and ClusterONE [81]. Additionally, the added value attained by merging procedures during postprocessing was assessed. Common protein complex prediction benchmarks as introduced in Section 2.1.2 were used in two evaluation steps including measures based either on the agreement with reference complexes and or on biological relevance.

The predicted sets of the algorithm developed in this thesis were taken without applying any filters, the output of the other methods was filtered for predicted candidates that contained at least one TF as annotated in the YPA (Section 3.2.4) to restrict the results to TF complexes. The results of the variants of the domain-aware cohesiveness optimization algorithm were obtained using the seed pairs as described earlier and a size threshold of 10 proteins to keep the effort low enough. Since the compared general approaches allow tuning of individual parameters, optimization of the most influential settings was conducted to obtain the most competitive individual and overall parameter sets for every single score. The best overall parameter set was determined by the achieved maximum of a combined score to which the averages of reference and biological scores contribute equally. Comprehensive results of this step are compiled in Appendix A.1.

MCODE was used as a Cytoscape3 plugin [138] in version 1.4b2 and run within Cytoscape3. Since MCODE can not handle weighted edges, the PrePPI-PPIN needed to be binarized using a certain threshold. Additionally, the clustering prediction process can be influenced by a node-score cutoff parameter. MCODE is able to predict overlapping complex candidates. Table A.1 summarizes the influence of the parameters.
For MCL the official standalone binary from http://micans.org/mcl/ in version 12-068 was used. An inflation parameter can be tuned to change the granularity of the predicted complexes. MCL can use weighted edges but does not allow for overlaps between complex candidates. The optimization results are given in Table A.2.
ClusterONE results were generated with the standalone implementation in version 1.0 from http://www.paccanarolab.org/cluster-one/. Due to completely different fundamental methodologies, ClusterONE is the only compared program that allows the user to manually influence seed nodes (or increase the importance of individual nodes) or sets of seed nodes. Both offered growth-start methods, in the following called seeded and unseeded, were utilized and the penalty parameter was optimized independently for all initialization variants. In the unseeded mode the implementation takes care of the initialization, but additionally two different case-specific seeded variants were set by hand: in one case, all YPA-annotated TFs were set as initial start proteins and in the other case, all pairs as used to initialize the domain-aware cohesiveness optimization were also used to commence the ClusterONE approach to also assess the benefit of an induced combinatorial flavor here as suggested in Section 3.1.1. Section 2.1.1 presented

the internals of ClusterONE in detail. Tables A.3 and A.4 in the Appendix show the detailed results of all three initialization variants.

Table 4.1 lists the quantities of predicted TF-containing complexes and distinct combinations of TFs that are predicted by the individual programs. For methods that underwent parameter optimization two results are shown for every measure: the results of the overall best performing parameter set per method are shown on the left and the best individual results achieved by any parameter set on the right.

	M	Mw	Cl1ps	Cl1s	Cl1	MCD	MCL
TF complexes	2331	1375	175/176	61/63	106/106	16/38	75/79
TF variants	699	412	134/138	59/61	80/80	16/38	75/79

Table 4.1: Overview of predicted TF complexes and TF complex variants (how many distinct combinations of TFs are involved in complexes) that were suggested by various approaches. The different methods are abbreviated as M (our method, state condensation variant), Mw (our method, weight-pruning variant), Cl1ps (ClusterONE, seeded from pairs), Cl1s (ClusterONE, seeded), Cl1 (ClusterONE, unseeded), MCD (MCODE) and MCL. If two numbers are shown, the left one represents the result of the overall best parameter set for the approach and the right one the highest value achieved by any parameter set. A shrinked version of the table can also be found in [1].

Our new algorithm in its two implementations suggested many more TF complex candidates than all the established methods. As expected, the ClusterONE run initialized with the pre-built pairs, as also used for the domain-aware method of the thesis, returned the second most TF complexes and variants. While MCL and ClusterONE, depending on the exact start setup, shared the range in between, MCODE placed as a distant last when it comes to the sheer quantity. Also, MCODE's output size deviated the most between overall best and individual best parameter set.

The differences between the start variants for ClusterONE, also locally optimizing the cohesiveness but without the domain model, were quite surprising: when initialized with the individual TFs as seeds - the protein neighborhood to locally aim for - fewer complexes were predicted as compared to just starting the growth from hub proteins in the PPIN, irrespective of the type of the hub proteins. Likewise worth mentioning is the comparably small number of candidates suggested by the growth with the 1526 pre-built pairs when compared to the two slightly different domain-aware approaches. Although initiated with seeds that should encourage the combinatorial manifold, the number of predicted TF variants was still even below the number of annotated TFs (148). This strengthens the early assumptions proposed in Section 3.1.1 where we argued that an approach that only uses protein interaction data is not sufficient to grasp the biological borders of such highly modular subnetworks as defined by, for example, binding-interface constraints. It seems likely that individual units were aggregated to a high degree since common complex prediction algorithms generally aim to delimit complexes that are fixed functional assemblies.

4.2.1   Comparison to reference complexes

After this first look at the number of predictions made by each method, the quality of the predictions needs to be assessed. Initially, established measures for complex prediction quality based on the agreement with reference datasets were deployed as described in Section 2.1.2.

All measures were independently evaluated on three different protein complex reference datasets for yeast, namely CYC2008, MIPS and SGD, as described in Section 3.2.6. With the exception of the precision, all quality metrics were calculated according to the corresponding reference complex sets filtered to the subset of known complexes that involve at least one YPA-annotated TF. The precision calculation was allowed to match a candidate to a reference complex that does not include a TF. This was done to facilitate matches to potentially recruited larger complexes of regulatory function in yeast, such as SWI/SNF or RSC. A method that predicts a candidate with high overlap to such a complex and contains at least one TF should definitively not be placed at a disadvantage. The threshold of overlap-score based tests was set to $\omega(A, B) > 0.25$ as done in other related publications [81, 93].Table 4.2 shows the results of the comparison-

based benchmarks for different reference datasets. On average, our novel approach(es) designed for this combinatorial protein complex prediction task outperformed the established general protein complex programs by a large margin. Especially MCODE, the program that predicted by far the smallest number of complexes, performed notably worse in comparison. Quite surprisingly, ClusterONE when started from the curated pairs was on

Ref.		M	Mw	Cl1ps	Cl1s	Cl1	MCD	MCL
	Prec	0.376	0.388	0.194/0.194	0.279/0.279	0.264/0.264	0.250/0.357	0.240/0.240
	Rec	0.786	0.964	0.357/0.357	0.357/0.357	0.357/0.393	0.179/0.179	0.429/0.500
CYC	FSc	0.508	0.554	0.252/0.252	0.313/0.313	0.304/0.309	0.208/0.208	0.308/0.319
	GeoA	0.762	0.785	0.738/0.740	0.777/0.783	0.768/0.773	0.663/0.663	0.744/0.775
	MMR	0.391	0.499	0.220/0.220	0.208/0.208	0.182/0.190	0.084/0.084	0.235/0.288
	Prec	0.244	0.201	0.109/0.114	0.131/0.131	0.132/0.151	0.125/0.143	0.093/0.101
	Rec	0.615	0.615	0.346/0.346	0.231/0.231	0.346/0.346	0.154/0.154	0.192/0.385
MIPS	FSc	0.349	0.304	0.165/0.172	0.167/0.171	0.191/0.210	0.138/0.148	0.126/0.160
	GeoA	0.382	0.369	0.535/0.537	0.427/0.433	0.537/0.547	0.372/0.397	0.392/0.516
	MMR	0.332	0.343	0.156/0.158	0.099/0.099	0.134/0.136	0.050/0.050	0.112/0.129
	Prec	0.325	0.311	0.194/0.195	0.213/0.230	0.198/0.198	0.125/0.167	0.160/0.192
	Rec	0.875	0.813	0.563/0.563	0.500/0.500	0.500/0.500	0.000/0.125	0.438/0.563
SGD	FSc	0.474	0.45	0.289/0.290	0.299/0.315	0.284/0.284	0.000/0.091	0.234/0.234
	GeoA	0.673	0.648	0.637/0.654	0.646/0.657	0.629/0.633	0.583/0.592	0.587/0.633
	MMR	0.417	0.458	0.220/0.220	0.205/0.205	0.210/0.213	0.041/0.073	0.193/0.237
	average	0.501	0.513	0.332/0.332	0.323/0.323	0.336/0.339	0.198/0.212	0.299/0.318

Table 4.2: Summary of the performance of complex prediction methods on reference complex data from CYC2008 [72], MIPS [130] and the SGD [183]. For the parameter-adjusted general methods each evaluation shows the result of the overall best parameter set on the left side and the best result achieved by any parameter set on the right side. Again, methods names are shortened to M (our method, state condensation variant), Mw (our method, weight-pruning variant), Cl1ps (ClusterONE, seeded from pairs), Cl1s (ClusterONE, seeded), Cl1 (ClusterONE, unseeded), MCD (MCODE) and MCL. The following short forms are used for brevity across all upcoming benchmarks based on the comparison with reference complexes: Prec: Precision (or FractionScore), Rec: Recall, FSc: F-score (or F-measure), GeoA: geometric accuracy and MMR: maximum matching ratio. The results can also be found in [1].

average not superior to the completely unsupervised version. Even the recall was only slightly better for one dataset while the precision was strictly lower. Only in the maximum matching ratio, due to its missing penalization of non-matching predictions, it was a slightly better. This showed that growing from pairs is not appreciably profitable for the local cohesiveness optimization as implemented by ClusterONE. This possibility was already theoretically discussed in Section 3.1.1 and attributed to the missing information of physical limitations in protein interaction networks. Therefore, ClusterONE tends to merge highly modular dense subnetworks (see Section 2.2).

	M	Mw	Cl1ps	Cl1s	Cl1	MCD	MCL
avg. size	9.4	6.7	14.6	13.2	18.9	91.0	16.0
avg. score	0.501	0.513	0.332	0.323	0.336	0.198	0.299

Table 4.3: Average sizes of predicted complexes related to average scores during reference complex assessments achieved per method. Methods names are shortened to M (proposed method, state condensation variant), Mw (proposed method, weight-pruning variant), Cl1ps (ClusterONE, seeded from pairs), Cl1s (ClusterONE, seeded), Cl1 (ClusterONE, unseeded), MCD (MCODE) and MCL. A shrinked version of the table can also be found in the supplement of [1].

Indeed, the average size of predicted complexes was smaller for the domain-aware methods, as can be seen in Table 4.3. When comparing the sizes to the average sizes of the reference complex datasets - CYC2008: about 4, MIPS: about 18, SGD: about 8 - the prediction of smaller complex candidates also seems to be advantageous; but, as the ClusterONE results show, smaller predictions were not necessarily better predictions. This would also mean a depth-threshold, as applied by the proposed methods and introduced for scalability, may be beneficial for this special clustering task. No method allowed to restrict the growth of its candidates - for most underlying algorithms it is actually not even technically possible, this affects MCODE, MCL and many other based on graph-theoretic foundations like spectral clustering [61]. Therefore, it could only be tested if a filtering of candidates to the ones including at most 10 proteins was already beneficial. This was tested on all ClusterONE variants, the outcome can be found in Table A.5 in the Appendix. Although for all initialization options most predicted complexes were within this size threshold anyhow, the quality measure only decreased when the complex size was restricted. So it was not possible to gain any task-specific advantage by an intervention by the user.

Moreover, another important observation is shown in Table 4.3. When the internal weight threshold was used, the average complex size was clearly decreased. Since this threshold is the only relevant difference between the two implementations with respect to the outcomes, the difference indicates that the local topology may actually not afford further connections and thus prevents a further growth based on the connectivity data. Since this happens naturally if the data suggests so, utilizing the weights is certainly a worthwhile advantage.

Recapitulating, for the intricate task of TF complex prediction the suggested approaches were clearly superior than common complex prediction programs when assessed using established quality measures based on the accordance with reference complexes. The incorporation of an internal probability threshold led to a better accordance when compared to the variant without this feature.

4.2.2 Assessment of biological relevance

Further quality measures - to complement and verify the previous benchmarks - rest upon the assumptions of colocalization and functional homogeneity within complexes (see Section 2.1.2). In the special case of TF complexes one should even expect an in-vivo localization at the nucleus for all proteins within the same complex.

Table 4.4 shows the results for those benchmarks.

Again, our novel approach(es) performed very well. MCODE, for example, delivered the best scoring candidates with respect to GO enrichment. However, when it was executed with a parameter set that also led to competitive agreement with reference complexes, on average less than half of the proteins within the predicted TF complexes were proteins that are actually also found in the nucleus. MCL was not even able to achieve more than 43% with any setting.

For the particular task of TF complex prediction, the nucleus colocalization score should be clearly seen as the most important one among the measures of biological relevance. Consequently, the domain-aware cohesiveness optimization approaches also succeed in this sector.

	M	Mw	Cl1ps	Cl1s	Cl1	MCD	MCL
NColoc	0.708	0.783	0.654/0.654	0.802/0.802	0.537/0.613	0.492/0.697	0.393/0.431
GOE	0.979	0.913	0.834/0.840	0.770/0.778	0.858/0.858	1.000/1.000	0.693/0.808
GOE(MF)	0.891	0.808	0.743/0.747	0.656/0.672	0.802/0.802	1.000/1.000	0.693/0.693
GOE(BP)	0.962	0.876	0.800/0.807	0.754/0.762	0.830/0.830	1.000/1.000	0.720/0.769
GOE(CC)	0.876	0.751	0.691/0.691	0.656/0.696	0.717/0.717	0.938/0.938	0.493/0.692
average	0.883	0.826	0.744/0.744	0.728/0.728	0.749/0.749	0.886/0.886	0.598/0.661

Table 4.4: Assessment of biological relevance for all evaluated approaches. Abbreviations used as before: M (our method, state condensation variant), Mw (our method, weight-pruning variant), Cl1s (ClusterONE, seeded), Cl1 (ClusterONE, unseeded), MCD (MCODE) and MCL. The following short forms are used for brevity across all upcoming benchmarks based on biological relevance: NColoc: nucleus colocalization, GOE: over-representation score based on GO term enrichment (all terms), GOE(MF): GO term enrichment (molecular function), GOE(BP): GO term enrichment (biological process) and GOE(CC): GO term enrichment (cellular component). If two numbers are shown, the left one represents the result of the overall best parameter set for the approach and the right one the highest value achieved by any parameter set. The results can also be found in [1].

4.2.3 Evaluation of postprocessing and thresholds

Merging highly overlapping complex candidates is an often applied post-processing step during protein complex prediction [81]. For the proposed algorithms it could be also of additional value to compensate for the restricted complex size originating from the depth-threshold. Three different methods were compared: a simple method based on the exact postprocessing as used by [81] and two versions of an iterative procedure that additionally imposes the DDI-model as a filter as done by [94]. The details were covered in Section 3.3.2.2.

Table 4.5 shows the individual results.

When objectively compared to the benchmark results achieved by the unaltered complex candidates, the additional merging procedures were not considerable as worthwhile. The scores slightly decreased with every applied postprocessing option, the unconstrained merging procedure as used by default in ClusterONE being the worst choice in terms of the established measures.

Ref.		M	Mw	CM	CMw	UM	UMw	wCM	wCMw
	TF complexes	2331	1375	1702	1247	1202	1148	1912	1264
	TF variants	699	412	613	411	512	403	656	411
CYC	Prec	0.376	0.388	0.340	0.346	0.269	0.302	0.339	0.354
	Rec	0.786	0.964	0.786	0.964	0.714	0.964	0.786	0.964
	FSc	0.508	0.554	0.475	0.510	0.391	0.460	0.474	0.518
	GeoA	0.762	0.785	0.780	0.784	0.788	0.781	0.769	0.784
	MMR	0.391	0.499	0.392	0.499	0.374	0.493	0.389	0.499
MIPS	Prec	0.244	0.201	0.200	0.177	0.141	0.156	0.220	0.179
	Rec	0.615	0.615	0.692	0.615	0.654	0.615	0.615	0.615
	FSc	0.349	0.304	0.311	0.275	0.233	0.249	0.324	0.277
	GeoA	0.382	0.369	0.397	0.373	0.404	0.369	0.395	0.373
	MMR	0.332	0.343	0.338	0.348	0.322	0.349	0.341	0.347
SGD	Prec	0.325	0.311	0.297	0.262	0.230	0.218	0.298	0.268
	Rec	0.875	0.813	0.875	0.813	0.875	0.813	0.875	0.813
	FSc	0.474	0.45	0.443	0.396	0.365	0.343	0.445	0.403
	GeoA	0.673	0.648	0.701	0.673	0.715	0.708	0.696	0.669
	MMR	0.417	0.458	0.423	0.463	0.381	0.448	0.423	0.463
	NColoc	0.708	0.783	0.706	0.762	0.706	0.739	0.704	0.765
	GOE	0.979	0.913	0.972	0.905	0.963	0.896	0.975	0.906
	GOE(MF)	0.891	0.808	0.887	0.804	0.883	0.795	0.894	0.806
	GOE(BP)	0.962	0.876	0.95	0.863	0.933	0.851	0.954	0.865
	GOE(CC)	0.876	0.751	0.848	0.726	0.815	0.704	0.856	0.729
	ref. avg.	0.501	0.513	0.497	0.500	0.457	0.485	0.493	0.502
	bio. avg.	0.883	0.826	0.873	0.812	0.860	0.797	0.877	0.814
	bal. avg.	0.692	0.670	0.685	0.656	0.659	0.641	0.685	0.658

Table 4.5: Results of added merging procedures. The different options in the columns alternately use the unweighted and the weighted results as their starting points, starting with the unmodified outputs. The abbreviations of the predictions can be read in the following way: M and Mw solely are used as before, M/Mw as a suffix indicate the input data, a U as a prefix abbreviates the unconstrained merging procedure, C the constrained DDI-model-version and wC the constraint version utilizing weights per interaction. The results of benchmarks based on the comparison with reference complexes are listed in row ref. avg., bio. avg. denotes the averaged results of biological relevance assessments and bal. avg. is a balanced average between all scores obtained from the two previous averages to balance both contributions equally weighted. Other short forms are used as introduced earlier.

However, one finding was noteworthy. The complex candidates predicted with the weight-utilizing algorithm appeared to be more robust against fusion of highly overlapping sets in terms of their TF variants. Independent of the merge implementation that was used, their number was barely affected. This showed that the thus predicted variants must feature a higher degree of exclusiveness that could only originate from the added thresholding.

Returning to an earlier and related question, the potentially better adaptability of the weight-threshold implementation in practice, the influence of the size-threshold was assessed for both variants. Figure 4.2 shows the behaviour of selected quality measures, the recall and the nucleus colocalization score, relative to changes of the depth-threshold for both variants of the approach. Other measures were left out since they were less suggestive. The recall - measuring the fraction of reference complexes that were sufficiently well approximated by any predicted candidate - indeed was decreased (by about 8% for the version of the algorithm without a probability-cutoff) when the depth-threshold was increased beyond a certain cluster size since very small reference complexes were not matched anymore. When the associated probability of the underlying spanning tree on the domain level was used as a cutoff, the recall progressed much smoother. This can only be explained by a sufficiently meaningful adaption to the local network topology and should thus lead to a better generalization in general. Furthermore, the nucleus colocalization was also found to be better adjusted.

To sum up this part of the analysis, merging procedures have not shown to be profitable for the problem in yeast. However, more evidence recommended the incorporation of the weight-threshold by default to obtain a better generalization. Therefore all further evaluation steps were conducted with the predicted set of TF complex candidates from the domain-aware cohesiveness optimization algorithm exploiting the weights from the PPIN to constrain the growth to the likely candidates.

4.3   Analysis in the transcription factor context

Every prediction method should have a practical utility. The prediction of TF complex candidates should in the best case be sufficient to provide a set of assemblies that allows to infer mutual target genes as well as an exerted mode of action for every complex. This will now be assessed for the proposed yeast complex candidates.

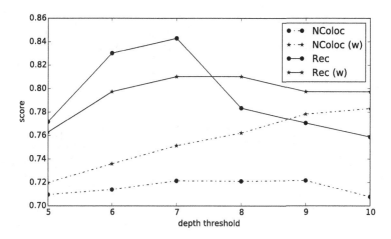

Figure 4.2: The nucleus colocalization and the recall (average recall across all three reference sets) were computed for the predictions of runs with depth-thresholds from five to ten for the version with the weight-threshold (scores with suffix (w)) and for the one without.

4.3.1 Estimation of target genes

The inference of target genes per predicted complex was consistently apportioned into four distinct classes with increasing restrictiveness by induction of constraints on pairwise distances between putative binding events as suggested by current models of cis-regulatory interplay (see Section 1.2.2). This in principle allows to have a more precise view on the set of potentially influenced genes and could later be worthwhile.

First all common targets, the set of mutual target genes of all involved TFs without any constraints, were determined. Also, common targets were considered that allowed for colocalization of the TFs - defined as binding regions where all TFs showed pairwise distances in the range of -50-50 bp, which means overlaps and corecruitment were conceded. The next two refinements had disjoint distance constraints. Targets of mediated cooperativity were defined as the genes that allowed for pairwise distances between 10-50 bp and targets with supposed direct cooperativity were strictly restricted to 0-10 bp.

Figure 4.3 shows the results of a target gene analysis where all predicted distinct TF combinations were tested according to the criteria above and additionally compared to systematic and sampled averages from all n-tuples of TFs. More and exact numbers are given in the Appendix A.2.

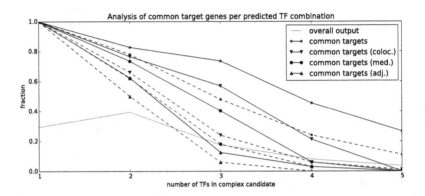

Figure 4.3: The plot shows different distributions related to n-tuples of TFs. All ratios were determined relative to the number of distinct n-tuples, since every common target calculation should only be taken into account once.
The gray marker-less line denotes the fraction of the respective n-tuples within the overall predicted distinct tuples while all other lines show the ratio of predicted TF tuples of a certain size with common target genes according to different binding site constraints (or none). All values were determined on the basis of all distinct TF combinations found together in complexes.
Dashed lines represent the correspondingly colored fractions of TF variants with a certain number of TFs when all permutations were systematically tested (pairs and triplets). The values for 4- and 5-tuples were determined as the average of 100,000 sampling experiments to approximate the proportions. The candidates that involved six (in total 6 candidates) and seven TFs (altogether 2 candidates) were left out for convenience. One of the predicted complexes with seven TFs had common target genes. Figure adapted from supplement of [1].

In total, 79% of the 413 distinct TF tuples (including single TFs) that were found within the predicted complexes shared common target genes. The predicted higher-order tuples consistently exhibited an above-average proportion of mutual target genes across all examined relevant n-tuples and biologically motivated constraints. This is especially noteworthy because complex prediction can only grasp cooperativity mediated by protein interactions while a complete sampling, as conducted for the 'baseline' measurement, can find all of them. In many cases the predicted set was 20%

above this average, for colocalized TF triple even 33% higher. Only the suggested 5-tuples were 1% lower than an average sampling when binding events of colocalization were examined. When evaluated for the objectively next best method, ClusterONE in the unseeded mode, pairs with common targets were 14% below the average, for example. Table A.8 in the Appendix shows all results for ClusterONE (unseeded).

Even though the information of TFs in complexes and their targets is already sufficient to automatically build a first gene regulatory network that includes potentially meaningful TF tuples, even more can be done with the output of the method since the recruited proteins have also been predicted.

4.3.2 Estimating the modes of action

In the ideal case the composition of proteins within every predicted TF complex allows to infer a distinct mode of action. This will now be evaluated for predicted yeast complexes on the example of a convenient and intuitive annotation-mining strategy.

Referring to the biological background in Section 1.2, TFs and TF complexes, roughly speaking, interact with the basal machinery or affect the chromatin structure and histone placement. The known epigenetic mechanisms in yeast are quite manageable. *S.cerevisiae* only exhibits proteins for histone acetylation and methylation of H3K4,36,79 whereas all modifications are basically associated with increased accessibility and therefore transcriptional upregulation [9]. Very few exceptions to this are known so far [191].
On the basis of this yeast-specific simplification we manually compiled a small list of potentially descriptive GO terms and retrieved the associated proteins using the QuickGO web service [192] during 16. Jan. 2014 (updated daily). Table 4.6 outlines the exact terms, elucidates their choice and their affiliation.

First, all predicted protein assemblies across all TF variants were checked for contradictory statements, e.g., if they harboured proteins that are marked to positively influence transcription as well as proteins that are annotated with repressory function. Only 17% of the predicted complexes comprised conflicting proteins.
Among the consistent candidates, 79% had at least one of the annotation

GO	initial	cleaned	name
GO:0045944	210	135	positive regulation of transcription from RNA polymerase II promoter
GO:0000122	125	54	negative regulation of transcription from RNA polymerase II promoter
GO:0003713	32	22	transcription coactivator activity
GO:0003714	16	6	transcription corepressor activity
GO:0004402	26	25	histone acetyltransferase activity
GO:0004407	15	9	histone deacetylase activity
GO:0042054	11	11	histone methyltransferase activity
GO:0032452	4	3	histone demethylase activity
GO:0006338	78	78	chromatin remodeling
GO:0016591	77	77	DNA-directed RNA polymerase II, holoenzyme

Table 4.6: The table shows the utilized set of intuitively chosen case-specific GO terms for yeast and the number of retrieved proteins falling into the category. While 'chromatin remodeling' and 'PolII holoenzyme', summarizing all proteins involved in the basal transcription machinery and the Mediator complex, can only supplement a binary indication of up- or downregulation, the other terms, roughly grouped into direct, cofactor-mediated and epigenetic contributions, also yield a distinct direction. Proteins that were annotated with both positive and negative influence were excluded to prevent the introduction of ambiguity due to context-dependent functions; the extent of the purification on the sizes of the sets is documented in the columns 'initial' and 'cleaned'. Table was also used in supplement of [1].

as considered, and for 65% even the direction of the contribution to the regulation could be inferred. For sole annotations of chromatin remodeling or PolII-holoenzyme such an association is not possible. About 3% of the TF combinations with at least one conflict-free complex candidate were part of assemblies with opposing regulatory effect.

RFX1/REB1, for example, is a pair of TFs that occurred in 10 slightly different predicted candidates. Four of those were annotated with activating proteins that are parts of the RSC complex, for two complexes repressory function was deduced and eight involved chromatin remodelling. All candidates were devoid of interactions with the basal machinery or contradictory annotations. Indeed, RFX1 is known to change its function by recruitment of the CYC8/TUP1 corepressor complex [193]. Both corepressor proteins were indeed found in the predicted repressive candidates and their GO annotations were actually the decisive factor for the according classification. Furthermore, REB1 is associated with nucleosome depleted regions, making the recruitment of remodellers seem plausible [194].

Another good example testifying the usefulness of the additional functional information inherent to the knowledge of an actual assembly, were the seven predicted complexes with FHL1 as the sole TF. Of those candidates two were annotated as activating, two as repressory, one had no inferable direction and one exhibited a conflict. Activating function of FHL1 is always accompanied by recruitment of coactivator IFH1 [195], this was also reflected in the predicted complexes. This shows why even for single TFs the knowledge of putative partners can be quite valuable.

While this strategy was only intended to exemplify a very simple toy approach that may be sufficient for yeast, for higher eukaryotes it is clearly not adequate. However, the potential information of the mode of action carried by the recruited proteins is a huge benefit in comparison to common information on regulatory networks. Some day, when computational models of transcription may even take the histone code into account, possible modifications could be inferred with equivalent level of detail from recruited writer-proteins.

4.3.3 Significance in yeast cell cycle

Until now all predictions could only provide a superordiate perspective. It is, strictly speaking, not self-evident that the proteins of a predicted complex are ever occurring together at the same time and space in vivo to carry out a certain control mechanism. Therefore, as a final test, all predicted pairs and higher-order tuples of TFs (290 in total) were submitted to a significance analysis of their targets' coexpression within the prepared cell cycle data (Section 3.2.5). Such an analysis is based on the idea that genes that are regulated by the same mechanism should exhibit a highly similar expression pattern (see Section 2.3.1).

The expression coherence scoring (ECS) threshold D was initialized as the average of $10,000$ repeats and every sampling experiment was based on $10,000$ replicates, making $p = 10^{-4}$ the highest achievable significance. Methodical details were provided in the corresponding Section 3.3.2.2.

At first the significance of dECS was calculated, a measure to assess how valuable the refinement of target genes is that is induced by the binding site constraints of an additional TF. 17 of the 290 predicted higher-order TF combinations led to a significantly higher ECS in the context of the cell cycle ($p < 0.05$). Subjected to GO term enrichment analysis, 76% of the corresponding target gene sets were significantly enriched with specific biological process annotations ($p < 0.05$). Table 4.7 summarizes the results for the dECS evaluation.

Unsurprisingly, all significant tuples were associated with either cell cycle control itself or metabolic processes that are in plausible crosstalk with the cell cycle. For many of the tuples literature evidence was found.

Combinatorial assemblies of MET4, MET31, MET32 and CBF1 are the key regulators of the sulfur metabolism in yeast [44]. For MET4/MET32 even the correct regulatory influence ($+$) was deducible for the two complexes containing the TFs; for the complex containing CBF1/MET32 no function could be annotated with the basic GO annotation strategy. This was actually also correct since MET4 recruitment is required to obtain a regulatory effect; CBF1 and MET32 are only responsible for the correct gene targeting. Also, colocalization was the most significant binding mode. This is supported by the fact that MET4 does not have a DNA-binding domain on its own and needs to be corecruited by the other TFs, thus could appear in an overlapping genomic location for valid binding events [44].

Figure 4.4 shows the complex-induced refinement in expression coherence among the target genes of MET4/MET32.

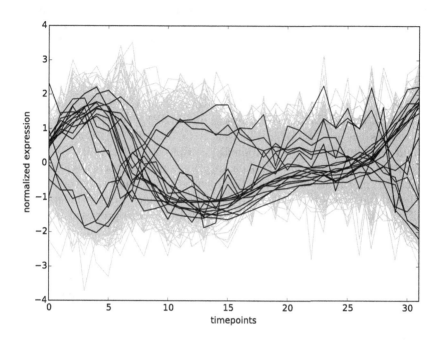

Figure 4.4: When the cell cycle expression profiles of all genes targeted by MET4 and MET32 (gray) are compared to the refined set of target genes where MET4 and MET32 bind as a colocalized complex (black), the increase in coherence is even truly significant with the bare eye. Figure was also used in [1].

DIG1/STE12 actually functions as a repressor of pheromone-responsive transcription as annotated in the corresponding predicted complex [196], the pair FHL1/RAP1 is a known regulator of the transcription of rRNA and ribosomal proteins and the three corresponding complex candidates were consistently associated with increase of transcription [197].
Furthermore, DAL80/GZF3 are dimerizing paralogs that indeed decrease the transcription of their target genes [183, 198], the GLN3/DAL80 and GLN3/GZF3 predictions are, however, artifacts due to the relatedness of their binding sites [199].
MSN2/MSN4 are paralogs responsible for stress response but nothing is

known about complex assembly [183], GCN4/RAP1 are known to cooperate at least indirectly by cooperative competition [200].

RPH1/GIS1 are indeed working together as repressors, but only during normal growth [201]. Since the predicted complexes are - at least to a certain degree - context-independent, the best possible prediction may have suggested at least one additional complex including RPH1 and GIS1, or at least one of the TFs, that could be associated with activation. But since this change of function is, at least in the case RPH1, related to post-translational modifications [202], it could by no means be detected by the approach. Moreover, the essential drivers of the G_1/S phase cell cycle transition, SWI4/SWI6 (SBF complex) and MBP1/SWI6 (MBF complex), as well as the pair STB1/SWI6 that is relevant for G_1/S as well were found reliably [203, 204]. Additionally, a triple of MBP1/SWI6/SWI4 was predicted. This did not come as a surprise, since the corresponding SBF and MBF motifs are indeed functionally overlapping and they cooccur significantly in promotors [116, 205]. Related to that, a cyclin-independent self-oscillatory circuit of the yeast cell cycle TFs was proposed based on binding data and tested with boolean regulatory network simulations. In a successful simulation, the SBF and MBF complexes were both needed for an activation of HCM1 and YOX1 (strict AND-logic) and only one of them for YHP1 (OR-logic) [206]. The conflict-free predicted complexes including MBP1/SWI6/SWI4 were all annotated as activating, and when a slightly expanded definition of mediated binding was used, namely pairwise distances of 0-50 bp, in fact HCM1 and YOX1 were in the set of target genes and YHP1 was not, matching the proposed AND-logic. In turn, when overlaps were allowed (as in colocalization), enabling SBF and MBF binding events in the same overlapping region, YHP1 was also in the set of target genes, supporting the OR-logic for YHP1. However, HCM1 and YOX1 were also targeted upon those constraints owed to the relaxed rules, refusing a clear differentiation. Although the domain-model admitted the TF triple, a combination of two heterodimers seems therefore more likely assuming decent binding data.

Since the evaluation of expression data conducted so far is known to be rather restrictive [122], additionally only the significance of the ECS itself was computed. In that respect, 37 of the TF tuples within predicted complexes were rated as significant ($p < 0.05$). Of their target sets, 78% were enriched in GO process annotations related to the cell cycle and elementary metabolism.

Besides many already mentioned combinations, even a significant quadruplet was detected that suggested colocalization of GCN4 in a complex with MET4, MET31 and MET32. Targets of that prediction where enriched with 'methionine metabolic process'. GCN4 is indeed known to bind to MET31 [65], but there is no literature evidence supporting an actual regulatory relationship. Other predicted pairs, like HAP4/HAP5 [207] and ASH1/DOT6 [208] were again backed by experimental findings.

TFs	p_{ABCS}	bind. mode	targets	reg. influence	GO process enrichment ($p < 0.05$) in targets
MET4/MET32	0.0010	coloc.	19	+	methionine metabolic process
TBP/HAP5	0.0335	med.	47	+	/
GLN3/DAL80	0.0009	med.	28	/	allantoin catabolic process
DIG1/STE12/SWI6	0.0369	all	15	/	fungal-type cell wall organization
FHL1/RAP1	0.0001	coloc.	116	+	rRNA transport
RPH1/GIS1	0.0001	med.	100	-	hexose catabolic process
CBF1/MET32	0.0002	coloc.	33	o	sulfate assimilation
DIG1/STE12	0.0003	med.	34	-	response to pheromone
GCN4/RAP1	0.033	med.	62	+	/
MSN4/MSN2	0.0021	med.	105	+	oligosaccharide biosynthetic process
DAL80/GZF3	0.0044	med.	20	-	purine nucleobase metabolic process
SWI6/SWI4	0.0039	med.	53	+	regulation of cyclin-dependent protein serine/threonine kinase activity
STB1/SWI6	0.0275	all	47	+	/
TBP/SWI6	0.0159	med.	14	+	/
GLN3/GZF3	0.0120	adj.	31	/	allantoin catabolic process
MBP1/SWI6/SWI4	0.0307	med.	18	+	regulation of cyclin-dependent protein serine/threonine kinase activity
MBP1/SWI6	0.0124	adj.	25	/	cell cycle process

Table 4.7: The list of predicted TF combinations with significant increase of expression coherence among their mutual targets comprised 15 pairs and 2 triples. The calculations were conducted for different conceivable modes of targetting (all shared target proteins, direct adjacency, mediated adjacency and colocalization) as defined earlier to have a finer picture of the possible target-gene sets. Only the most enriched GO process term is shown for each target set. The inferred regulatory influence on the rate of transcription is abbreviated as follows: + (increase), - (decrease), o (no statement possible), / (conflicting annotations). Table was also used in [1].

Conclusion and Outlook

5.1 Conclusion

TF complexes are highly modular combinatorial complex assemblies and thus clearly different from large self-contained functional protein complexes. The novel methodology presented in the thesis was shown to be superior to common complex prediction programs for this sophisticated task of TF complex prediction in yeast.

More importantly, the proposed expectations regarding the eventual information content of the predictions were comprehensively satisfied. Many of the individual predictions could be associated with target genes and a potential regulatory effect. The combinatorial notion of the algorithm even allowed to detect assemblies that included the same set of TFs but exerted opposite regulatory function. Those findings were backed by literature evidence.

In addition, it was shown how predicted candidates could be identified as regulatory drivers during a defined cellular state and condition. By employing expression coherence analysis among their putatively regulated genes for cell cycle expression data, many known complexes were obtained.

5.2 Outlook

The knowledge of TF complex candidates - their composition of DNA-binding members, but also the recruited potentially regulatory active proteins - offers extensive and, most of all, novel capabilities in the modeling of regulatory networks.

The discrete association of TF tuples to their target genes in a fully automated manner could replace handcrafted logical rules and tremendously simplify the construction of various types of regulatory models. This directly concerns the rulesets of boolean network simulations, all kinds of stochastic simulations but would also be applicable in statistical models by introducing sophisticated complexome basis functions.

Independently, the almost arbitrarily granular knowledge of the function of a recruited protein enables to exploit regulatory systems to yet unknown detail. It would, for example, be possible to model even single histone modifications in simulations, because the catalytic influence of corecruited proteins can be inferred from, for example, GO annotations.

The results and the broad range of applications suggest a deployment in higher eukaryotes, where the combinatorial interplay between transcription factors is thought to be more pronounced. However, there is certainly room for prior improvement and verification. Since the addition of an internal weight-threshold has been shown to be beneficial during the analyses, this poses the question if, for instance, a node-penalty (as used by ClusterONE) would further encourage the adaptability to the local topology.

Additional tables

A.1 Supplement to complex prediction benchmarks

Abbreviations: NPred: number of predicted candidates for TF complexes, NTF: number of distinct TF combinations among predicted complexes, Prec: Precision (or FractionScore), Rec: Recall, FSc: F-score (or F-measure), GeoA: geometric accuracy, MMR: maximum matching ratio, NColoc: nucleus colocalization, GOE: overrepresentation score based on GO term enrichment (all terms), GOE(MF): GO term enrichment (molecular function), GOE(BP): GO term enrichment (biological process), GOE(CC): GO term enrichment (cellular component), ref. avg.: average score when compared to reference complexes, bio. avg.: average score in biological relevance assessments, bal. avg.: balanced average between ref. and bio. averages (reference comparison and biological relevance weighted equally).

Ref.		e = 0.2					e = 0.3					e = 0.5					e = 0.7					e = 0.9				
		c=0.1	c=0.2	c=0.3	c=0.4	c=0.5	c=0.1	c=0.2	c=0.3	c=0.4	c=0.5	c=0.1	c=0.2	c=0.3	c=0.4	c=0.5	c=0.1	c=0.2	c=0.3	c=0.4	c=0.5	c=0.1	c=0.2	c=0.3	c=0.4	c=0.5
	NPred	34	27	25	21	16	38	31	24	17	14	21	28	22	19	12	15	17	18	17	15	6	16	16	16	14
	NTF	34	27	25	21	16	38	31	24	17	14	21	28	22	19	12	15	17	18	17	15	6	16	16	16	14
CYC	Prec	0.029	0.037	0.080	0.000	0.000	0.053	0.065	0.083	0.059	0.071	0.095	0.000	0.000	0.053	0.000	0.000	0.000	0.056	0.235	0.200	0.333	0.063	0.125	0.250	0.357
	Rec	0.000	0.000	0.000	0.000	0.000	0.036	0.036	0.000	0.036	0.036	0.036	0.000	0.000	0.000	0.000	0.000	0.000	0.036	0.107	0.107	0.071	0.036	0.036	0.179	0.143
	FSc	0.000	0.000	0.000	0.000	0.000	0.043	0.046	0.000	0.044	0.048	0.052	0.000	0.000	0.000	0.000	0.000	0.000	0.043	0.147	0.140	0.118	0.045	0.056	0.208	0.204
	GeoA	0.517	0.494	0.495	0.487	0.405	0.488	0.468	0.488	0.466	0.422	0.472	0.302	0.462	0.321	0.444	0.458	0.481	0.488	0.505	0.492	0.327	0.615	0.606	0.663	0.638
	MMR	0.033	0.025	0.014	0.005	0.002	0.069	0.034	0.011	0.012	0.014	0.051	0.028	0.022	0.029	0.011	0.024	0.025	0.030	0.064	0.058	0.034	0.040	0.047	0.084	0.054
MIPS	Prec	0.059	0.000	0.000	0.000	0.000	0.026	0.000	0.000	0.000	0.000	0.000	0.000	0.000	0.000	0.000	0.000	0.000	0.056	0.059	0.067	0.167	0.000	0.125	0.125	0.143
	Rec	0.077	0.000	0.000	0.000	0.000	0.038	0.000	0.000	0.000	0.000	0.000	0.000	0.000	0.000	0.000	0.000	0.000	0.038	0.038	0.038	0.038	0.000	0.038	0.154	0.154
	FSc	0.067	0.000	0.000	0.000	0.000	0.031	0.000	0.000	0.000	0.000	0.000	0.000	0.000	0.000	0.000	0.000	0.000	0.045	0.047	0.049	0.063	0.000	0.059	0.138	0.148
	GeoA	0.369	0.403	0.409	0.432	0.425	0.340	0.372	0.387	0.397	0.394	0.260	0.329	0.316	0.356	0.345	0.289	0.318	0.316	0.384	0.336	0.140	0.254	0.332	0.372	0.384
	MMR	0.039	0.021	0.011	0.010	0.004	0.027	0.022	0.011	0.006	0.003	0.013	0.019	0.019	0.015	0.003	0.014	0.022	0.022	0.022	0.020	0.030	0.014	0.046	0.059	0.037
SGD	Prec	0.000	0.037	0.040	0.000	0.000	0.026	0.032	0.083	0.000	0.000	0.048	0.000	0.000	0.000	0.000	0.000	0.000	0.059	0.061	0.065	0.091	0.000	0.083	0.000	0.087
	Rec	0.000	0.000	0.000	0.000	0.000	0.125	0.000	0.000	0.000	0.000	0.000	0.000	0.000	0.000	0.000	0.000	0.000	0.063	0.063	0.063	0.091	0.000	0.083	0.000	0.087
	FSc	0.443	0.414	0.423	0.423	0.417	0.417	0.394	0.452	0.457	0.411	0.411	0.462	0.475	0.423	0.448	0.456	0.448	0.468	0.467	0.438	0.251	0.530	0.538	0.583	0.592
	GeoA	0.443	0.414	0.423	0.423	0.417	0.414	0.394	0.452	0.457	0.411	0.411	0.462	0.475	0.423	0.448	0.456	0.448	0.468	0.467	0.438	0.251	0.530	0.538	0.583	0.592
	MMR	0.026	0.009	0.008	0.006	0.002	0.064	0.015	0.007	0.004	0.002	0.028	0.010	0.008	0.002	0.002	0.022	0.017	0.034	0.031	0.026	0.024	0.028	0.073	0.041	0.030
	NColoc	0.390	0.425	0.414	0.417	0.391	0.420	0.412	0.411	0.408	0.398	0.555	0.474	0.492	0.427	0.429	0.608	0.518	0.513	0.445	0.444	0.697	0.585	0.543	0.492	0.463
	GOE	0.882	0.889	0.960	0.952	0.750	0.968	0.903	0.875	1.000	1.000	0.952	0.929	0.955	0.947	0.917	0.867	0.882	1.000	1.000	1.000	0.833	0.875	0.938	1.000	1.000
	GOE(MF)	0.676	0.815	0.810	0.810	0.813	0.737	0.871	0.792	0.882	0.857	0.857	0.882	0.818	0.842	0.917	0.867	0.882	0.944	0.882	0.933	1.000	0.938	0.938	0.886	0.929
	GOE(BP)	0.853	0.889	0.920	0.952	0.750	0.842	0.903	0.833	0.941	1.000	0.905	0.893	0.955	0.947	0.917	0.933	0.882	1.000	0.882	0.933	0.833	0.813	0.938	1.000	0.929
	GOE(CC)	0.735	0.741	0.760	0.762	0.813	0.868	0.903	0.792	0.588	0.714	0.667	0.714	0.818	0.737	0.750	0.733	0.706	0.722	0.765	0.800	0.667	0.813	0.938	0.938	0.929
	ref. avg.	0.110	0.096	0.069	0.091	0.084	0.122	0.099	0.101	0.099	0.093	0.098	0.090	0.088	0.063	0.084	0.080	0.087	0.121	0.153	0.144	0.128	0.113	0.157	0.196	0.212
	bio. avg.	0.707	0.752	0.795	0.779	0.703	0.747	0.798	0.741	0.764	0.794	0.787	0.773	0.790	0.780	0.786	0.802	0.774	0.836	0.795	0.822	0.806	0.805	0.859	0.886	0.850
	bal. avg.	0.408	0.424	0.447	0.435	0.393	0.434	0.449	0.421	0.431	0.444	0.443	0.431	0.439	0.436	0.435	0.441	0.430	0.478	0.474	0.483	0.467	0.459	0.508	0.542	0.531

Table A.1: Parameter tuning for MCODE involves the network edge cutoff e to binarize the weighted network and the node score cutoff c. Based on the balanced average score between reference complexes and biological significance, the optimal values for network and reference data are $e = 0.9$ and $c = 0.4$. Table was also used in supplement of [1].

Ref.		$i = 1.5$	$i = 2.0$	$i = 2.5$	$i = 3.0$	$i = 3.5$	$i = 4.0$	$i = 4.5$	$i = 5.0$
	NPred	26	53	71	75	75	79	74	70
	NTF	26	53	71	75	75	79	74	70
	Prec	0.115	0.170	0.183	0.240	0.227	0.203	0.189	0.129
	Rec	0.000	0.250	0.393	0.429	0.536	0.500	0.500	0.321
CYC	FSc	0.000	0.202	0.250	0.308	0.319	0.288	0.275	0.184
	GeoA	0.432	0.763	0.775	0.744	0.727	0.701	0.670	0.596
	MMR	0.006	0.154	0.246	0.235	0.288	0.273	0.265	0.207
	Prec	0.038	0.057	0.099	0.093	0.093	0.101	0.095	0.100
	Rec	0.000	0.077	0.269	0.192	0.346	0.385	0.346	0.346
MIPS	FSc	0.000	0.065	0.144	0.126	0.147	0.160	0.149	0.155
	GeoA	0.410	0.516	0.424	0.392	0.269	0.259	0.248	0.241
	MMR	0.008	0.080	0.129	0.112	0.129	0.117	0.115	0.107
	Prec	0.192	0.075	0.099	0.160	0.133	0.139	0.122	0.100
	Rec	0.000	0.188	0.313	0.438	0.563	0.563	0.563	0.438
SGD	FSc	0.000	0.108	0.150	0.234	0.216	0.223	0.200	0.163
	GeoA	0.436	0.615	0.633	0.587	0.554	0.533	0.515	0.459
	MMR	0.004	0.122	0.226	0.193	0.223	0.233	0.237	0.169
	NColoc	0.419	0.431	0.408	0.393	0.384	0.398	0.414	0.413
	GOE	0.808	0.698	0.676	0.693	0.653	0.620	0.595	0.586
	GOE(MF)	0.615	0.623	0.662	0.693	0.680	0.608	0.581	0.557
	GOE(BP)	0.769	0.698	0.676	0.720	0.653	0.633	0.581	0.571
	GOE(CC)	0.692	0.509	0.479	0.493	0.440	0.443	0.392	0.371
	ref. avg.	0.109	0.229	0.289	0.299	0.318	0.312	0.299	0.248
	bio. avg.	0.661	0.592	0.580	0.598	0.562	0.540	0.513	0.500
	bal. avg.	0.385	0.410	0.434	0.449	0.440	0.426	0.406	0.374

Table A.2: Parameter optimization for MCL involves the inflation parameter i. Based on the balanced average, an inflation of $i = 3.0$ is suggested. Table was also used in supplement of [1].

		unseeded					seeded									
Ref.		$p=0.0$	$p=0.5$	$p=1.0$	$p=1.5$	$p=2.0$	$p=0.0$	$p=0.5$	$p=1.0$	$p=1.5$	$p=2.0$	$p=2.5$	$p=3.0$	$p=3.5$	$p=4.0$	$p=4.5$
	NPred	101	106	106	104	106	56	56	53	60	60	59	61	63	61	61
	NTF	78	79	80	77	77	56	55	51	57	57	57	59	61	59	59
CYC	Prec	0.208	0.255	0.264	0.240	0.226	0.250	0.268	0.264	0.250	0.267	0.271	0.262	0.254	0.279	0.246
	Rec	0.357	0.393	0.357	0.321	0.321	0.286	0.321	0.321	0.321	0.357	0.357	0.357	0.357	0.357	0.250
	FSc	0.263	0.309	0.304	0.275	0.266	0.267	0.292	0.290	0.281	0.305	0.308	0.302	0.297	0.313	0.248
	GeoA	0.748	0.760	0.768	0.767	0.773	0.637	0.632	0.628	0.683	0.695	0.783	0.773	0.778	0.777	0.783
	MMR	0.190	0.189	0.182	0.176	0.176	0.163	0.172	0.173	0.172	0.178	0.190	0.194	0.198	0.208	0.185
MIPS	Prec	0.149	0.151	0.132	0.135	0.142	0.107	0.107	0.113	0.117	0.117	0.136	0.131	0.127	0.131	0.131
	Rec	0.346	0.346	0.346	0.308	0.308	0.192	0.192	0.192	0.231	0.231	0.231	0.231	0.231	0.231	0.115
	FSc	0.208	0.210	0.191	0.187	0.194	0.138	0.138	0.143	0.155	0.155	0.171	0.167	0.164	0.167	0.123
	geoA	0.532	0.530	0.537	0.535	0.547	0.264	0.260	0.260	0.402	0.401	0.432	0.426	0.428	0.427	0.433
	MMR	0.136	0.133	0.134	0.119	0.121	0.072	0.072	0.072	0.088	0.088	0.094	0.099	0.099	0.099	0.085
SGD	Prec	0.198	0.189	0.198	0.173	0.179	0.214	0.214	0.226	0.200	0.200	0.220	0.213	0.206	0.213	0.230
	Rec	0.500	0.500	0.500	0.500	0.500	0.500	0.500	0.500	0.500	0.500	0.500	0.500	0.500	0.500	0.500
	FSc	0.284	0.274	0.284	0.257	0.264	0.300	0.300	0.312	0.286	0.286	0.306	0.299	0.292	0.299	0.315
	GeoA	0.629	0.631	0.629	0.628	0.633	0.502	0.504	0.497	0.552	0.550	0.657	0.653	0.646	0.646	0.646
	MMR	0.211	0.213	0.210	0.208	0.209	0.149	0.158	0.169	0.164	0.167	0.186	0.187	0.189	0.205	0.205
	NColoc	0.613	0.571	0.537	0.576	0.545	0.646	0.632	0.632	0.702	0.710	0.797	0.801	0.798	0.802	0.772
	GOE	0.842	0.840	0.858	0.827	0.802	0.768	0.750	0.736	0.750	0.750	0.763	0.770	0.778	0.770	0.754
	GOE(MF)	0.743	0.764	0.802	0.798	0.783	0.661	0.643	0.660	0.650	0.650	0.661	0.672	0.667	0.656	0.656
	GOE(BP)	0.812	0.802	0.830	0.798	0.774	0.732	0.714	0.717	0.733	0.733	0.746	0.754	0.762	0.754	0.738
	GOE(CC)	0.703	0.708	0.717	0.654	0.660	0.696	0.696	0.660	0.650	0.633	0.627	0.639	0.635	0.656	0.672
	ref. avg.	0.331	0.339	0.336	0.322	0.324	0.269	0.275	0.277	0.293	0.300	0.323	0.320	0.318	0.323	0.300
	bio. avg.	0.743	0.737	0.749	0.731	0.713	0.701	0.687	0.681	0.697	0.695	0.719	0.727	0.728	0.728	0.718
	bal. avg.	0.537	0.538	0.542	0.526	0.518	0.485	0.481	0.479	0.495	0.497	0.521	0.523	0.523	0.525	0.509

Table A.3: Parameter optimization for ClusterONE's penalty parameter p for the standard and the seeded runs independently. The best balanced average score is achieved with $p = 1.0$ for the unseeded and $p = 4.0$ for the seeded measurements. Table was also used in supplement of [1].

Ref.		$p = 0.0$	$p = 0.5$	$p = 1.0$	$p = 1.5$	$p = 2.0$
	NPred	175	175	174	176	169
	NTF	138	134	134	134	129
	Prec	0.183	0.194	0.184	0.165	0.154
	Rec	0.321	0.357	0.357	0.357	0.357
CYC	FSc	0.233	0.252	0.243	0.226	0.215
	GeoA	0.740	0.738	0.727	0.719	0.718
	MMR	0.209	0.220	0.220	0.217	0.218
	Prec	0.114	0.109	0.109	0.102	0.101
	Rec	0.346	0.346	0.346	0.308	0.308
MIPS	FSc	0.172	0.165	0.166	0.154	0.152
	GeoA	0.529	0.535	0.537	0.536	0.537
	MMR	0.155	0.156	0.158	0.140	0.142
	Prec	0.194	0.194	0.195	0.193	0.183
	Rec	0.563	0.563	0.563	0.563	0.563
SGD	FSc	0.289	0.289	0.290	0.288	0.277
	GeoA	0.621	0.637	0.640	0.654	0.652
	MMR	0.213	0.220	0.206	0.205	0.207
	NColoc	0.638	0.654	0.654	0.614	0.616
	GOE	0.840	0.834	0.833	0.841	0.828
	GOE(MF)	0.743	0.743	0.747	0.750	0.740
	GOE(BP)	0.806	0.800	0.799	0.807	0.805
	GOE(CC)	0.691	0.691	0.678	0.682	0.675
	ref. avg.	0.325	0.332	0.329	0.322	0.319
	bio. avg.	0.744	0.744	0.742	0.739	0.733
	bal. avg.	0.534	0.538	0.535	0.530	0.526

Table A.4: Parameter optimization of ClusterONE penalty p when starting with seeded pairs. Based on the results, $p = 0.5$ is the best choice in terms of all averaged measurements. Table was also used in supplement of [1].

Ref.		Cl1ps	Cl1s	Cl1
	NPred	105	42	63
CYC	Prec	0.152	0.286	0.222
	Rec	0.321	0.321	0.321
	FSc	0.207	0.303	0.263
MIPS	Prec	0.029	0.048	0.048
	Rec	0.154	0.154	0.154
	FSc	0.048	0.073	0.048
SGD	Prec	0.133	0.167	0.159
	Rec	0.375	0.313	0.313
	FSc	0.197	0.217	0.211

Table A.5: Results for the parameter optimized variants of ClusterONE filtered to include only complexes with at most 10 proteins. The results were restricted to the overlapscore-based measures since they are most affected by oversized estimation of complexes.

A.2   Supplement to target analysis

tuple size	all	1	2	3	4	5	6	7
number	413	121	162	72	33	15	6	2
contribution	1.00	0.29	0.39	0.17	0.08	0.04	0.01	0.00
common targets	0.79	1.0	0.83	0.74	0.45	0.27	0.00	0.50
common targets (coloc.)	0.71	1.00	0.77	0.57	0.21	0.00	0.00	0.00
common targets (med.)	0.66	1.00	0.73	0.40	0.06	0.00	0.00	0.00
common targets (adj.)	0.56	1.00	0.62	0.13	0.03	0.00	0.00	0.00

Table A.6: Target analysis of distinct TF tuples from the predictions. Definitions can be found in the corresponding continuous text.

tuple size	2	3	4	5
number	10, 878	529, 396	100, 000	100, 000
common targets	0.78	0.48	0.24	0.11
common targets (coloc.)	0.66	0.24	0.06	0.01
common targets (med.)	0.62	0.18	0.03	0.00
common targets (adj.)	0.49	0.06	0.00	0.00

Table A.7: Target analysis of distinct TF tuples from the systematic evaluations of all tuples for pairs and triplets and random sampling (100.000 repeats) for higher-order tuples. Definitions can be found in the corresponding continuous text.

tuple size	0	1	2	3	4	5	6	8	9	10	11
number	80	48	11	5	3	5	2	2	2	1	1
common targets	0.8	1.0	0.64	1.0	1.0	0.2	0.0	0.0	0.0	0.0	0.0
common targets (coloc.)	0.74	1.0	0.55	0.6	0.67	0.0	0.0	0.0	0.0	0.0	0.0
common targets (med.)	0.71	1.0	0.55	0.6	0.0	0.0	0.0	0.0	0.0	0.0	0.0
common targets (adj.)	0.69	1.0	0.55	0.2	0.0	0.0	0.0	0.0	0.0	0.0	0.0

Table A.8: Target analysis of distinct TF tuples from the ClusterONE predictions. Definitions can be found in the corresponding continuous text.

Bibliography

[1] T. Will and V. Helms. Identifying transcription factor complexes and their roles. *Bioinformatics*, 30(17):i415–i421, Sep 2014.

[2] B. Alberts. *Molecular Biology of the Cell*. Molecular Biology of the Cell. Garland Science, 5th ed. edition, 2008.

[3] Eric H. Davidson. *The Regulatory Genome: Gene Regulatory Networks In Development And Evolution*. Academic Press, 1st edition, 2006.

[4] G. A. Wray, M. W. Hahn, E. Abouheif, J. P. Balhoff, M. Pizer, M. V. Rockman, L. A. Romano, and G. A. Wray. The evolution of transcriptional regulation in eukaryotes. *Mol. Biol. Evol.*, 20(9):1377–1419, Sep 2003.

[5] T. Phillips and L. Hoopes. Transcription factors and transcriptional control in eukaryotic cells. *Nature Education*, 1(1):119, 2008.

[6] M. Levine and R. Tjian. Transcription regulation and animal diversity. *Nature*, 424(6945):147–151, Jul 2003.

[7] F. Crick. Central dogma of molecular biology. *Nature*, 227(5258):561–563, Aug 1970.

[8] D. Latchman and D.S. Latchman. *Eukaryotic Transcription Factors*. Eukaryotic Transcription Factors Series. Elsevier Science, 2010.

[9] C.D. Allis, T. Jenuwein, and D. Reinberg. *Epigenetics*. Cold Spring Harbor Laboratory Press, 1st edition, 2006.

[10] B. D. Strahl and C. D. Allis. The language of covalent histone modifications. *Nature*, 403(6765):41–45, Jan 2000.

[11] T. B. Miranda and P. A. Jones. DNA methylation: the nuts and bolts of repression. *J. Cell. Physiol.*, 213(2):384–390, Nov 2007.

[12] J. Teles, C. Pina, P. Eden, M. Ohlsson, T. Enver, and C. Peterson. Transcriptional regulation of lineage commitment–a stochastic model of cell fate decisions. *PLoS Comput. Biol.*, 9(8):e1003197, 2013.

[13] K. Hochedlinger and K. Plath. Epigenetic reprogramming and induced pluripotency. *Development*, 136(4):509–523, Feb 2009.

[14] K. M. Jozwik and J. S. Carroll. Pioneer factors in hormone-dependent cancers. *Nat. Rev. Cancer*, 12(6):381–385, Jun 2012.

[15] R.A. Weinberg. *The Biology of Cancer*. Number Bd. 1 in The Biology of Cancer. Garland Science, 2007.

[16] V. Pelechano, J. Garcia-Martinez, and J. E. Perez-Ortin. A genomic study of the inter-ORF distances in Saccharomyces cerevisiae. *Yeast*, 23(9):689–699, Jul 2006.

[17] S. Lubliner, L. Keren, and E. Segal. Sequence features of yeast and human core promoters that are predictive of maximal promoter activity. *Nucleic Acids Res.*, 41(11):5569–5581, Jun 2013.

[18] G. A. Maston, S. K. Evans, and M. R. Green. Transcriptional regulatory elements in the human genome. *Annu Rev Genomics Hum Genet*, 7:29–59, 2006.

[19] K. Murakami, H. Elmlund, N. Kalisman, D. A. Bushnell, C. M. Adams, M. Azubel, D. Elmlund, Y. Levi-Kalisman, X. Liu, B. J. Gibbons, M. Levitt, and R. D. Kornberg. Architecture of an RNA polymerase II transcription pre-initiation complex. *Science*, 342(6159):1238724, Nov 2013.

[20] F. Spitz and E. E. Furlong. Transcription factors: from enhancer binding to developmental control. *Nat. Rev. Genet.*, 13(9):613–626, Sep 2012.

[21] A. Z. Ansari and A. K. Mapp. Modular design of artificial transcription factors. *Curr Opin Chem Biol*, 6(6):765–772, Dec 2002.

[22] D. S. Latchman. Transcription factors: an overview. *Int. J. Biochem. Cell Biol.*, 29(12):1305–1312, Dec 1997.

[23] M. Pellegrini-Calace and J. M. Thornton. Detecting DNA-binding helix-turn-helix structural motifs using sequence and structure information. *Nucleic Acids Res.*, 33(7):2129–2140, 2005.

[24] G. Suske. The Sp-family of transcription factors. *Gene*, 238(2):291–300, Oct 1999.

[25] U. Schaefer, S. Schmeier, and V. B. Bajic. TcoF-DB: dragon database for human transcription co-factors and transcription factor interacting proteins. *Nucleic Acids Res.*, 39(Database issue):D106–110, Jan 2011.

[26] K. M. Lelli, M. Slattery, and R. S. Mann. Disentangling the many layers of eukaryotic transcriptional regulation. *Annu. Rev. Genet.*, 46:43–68, 2012.

[27] Z. C. Poss, C. C. Ebmeier, and D. J. Taatjes. The Mediator complex and transcription regulation. *Crit. Rev. Biochem. Mol. Biol.*, Oct 2013.

[28] D. Kadosh and K. Struhl. Targeted recruitment of the Sin3-Rpd3 histone deacetylase complex generates a highly localized domain of repressed chromatin in vivo. *Mol. Cell. Biol.*, 18(9):5121–5127, Sep 1998.

[29] A. K. Mal. Histone methyltransferase Suv39h1 represses MyoD-stimulated myogenic differentiation. *EMBO J.*, 25(14):3323–3334, Jul 2006.

[30] P. A. Jones and D. Takai. The role of DNA methylation in mammalian epigenetics. *Science*, 293(5532):1068–1070, Aug 2001.

[31] A. Rottach, H. Leonhardt, and F. Spada. DNA methylation-mediated epigenetic control. *J. Cell. Biochem.*, 108(1):43–51, Sep 2009.

[32] X. Liu, M. Luo, and K. Wu. Epigenetic interplay of histone modifications and DNA methylation mediated by HDA6. *Plant Signal Behav*, 7(6):633–635, Jun 2012.

[33] D. Thanos and T. Maniatis. Virus induction of human IFN beta gene expression requires the assembly of an enhanceosome. *Cell*, 83(7):1091–1100, Dec 1995.

[34] D. Panne. The enhanceosome. *Curr. Opin. Struct. Biol.*, 18(2):236–242, Apr 2008.

[35] D. Panne, T. Maniatis, and S. C. Harrison. An atomic model of the interferon-beta enhanceosome. *Cell*, 129(6):1111–1123, Jun 2007.

[36] M. Kazemian, H. Pham, S. A. Wolfe, M. H. Brodsky, and S. Sinha. Widespread evidence of cooperative DNA binding by transcription factors in Drosophila development. *Nucleic Acids Res.*, 41(17):8237–8252, Sep 2013.

[37] R. H. Goodman and S. Smolik. CBP/p300 in cell growth, transformation, and development. *Genes Dev.*, 14(13):1553–1577, Jul 2000.

[38] D. N. Arnosti and M. M. Kulkarni. Transcriptional enhancers: Intelligent enhanceosomes or flexible billboards? *J. Cell. Biochem.*, 94(5):890–898, Apr 2005.

[39] S. Lomvardas and D. Thanos. Modifying gene expression programs by altering core promoter chromatin architecture. *Cell*, 110(2):261–271, Jul 2002.

[40] P. M. Gowri, J. H. Yu, A. Shaufl, M. A. Sperling, and R. K. Menon. Recruitment of a repressosome complex at the growth hormone receptor promoter and its potential role in diabetic nephropathy. *Mol. Cell. Biol.*, 23(3):815–825, Feb 2003.

[41] M. M. Kulkarni and D. N. Arnosti. Information display by transcriptional enhancers. *Development*, 130(26):6569–6575, Dec 2003.

[42] G. Junion, M. Spivakov, C. Girardot, M. Braun, E. H. Gustafson, E. Birney, and E. E. Furlong. A transcription factor collective defines cardiac cell fate and reflects lineage history. *Cell*, 148(3):473–486, Feb 2012.

[43] T. Siggers, M. H. Duyzend, J. Reddy, S. Khan, and M. L. Bulyk. Non-DNA-binding cofactors enhance DNA-binding specificity of a transcriptional regulatory complex. *Mol. Syst. Biol.*, 7:555, 2011.

[44] T. A. Lee, P. Jorgensen, A. L. Bognar, C. Peyraud, D. Thomas, and M. Tyers. Dissection of combinatorial control by the Met4 transcriptional complex. *Mol. Biol. Cell*, 21(3):456–469, Feb 2010.

[45] M. I. Stefan and N. Le Novere. Cooperative binding. *PLoS Comput. Biol.*, 9(6):e1003106, 2013.

[46] G. Wedler. *Lehrbuch der Physikalischen Chemie*. Wiley, 5th edition, 2004.

[47] L. Giorgetti, T. Siggers, G. Tiana, G. Caprara, S. Notarbartolo, T. Corona, M. Pasparakis, P. Milani, M. L. Bulyk, and G. Natoli. Noncooperative interactions between transcription factors and clustered DNA binding sites enable graded transcriptional responses to environmental inputs. *Mol. Cell*, 37(3):418–428, Feb 2010.

[48] D. Aguilar and B. Oliva. Topological comparison of methods for predicting transcriptional cooperativity in yeast. *BMC Genomics*, 9:137, 2008.

[49] J. A. Miller and J. Widom. Collaborative competition mechanism for gene activation in vivo. *Mol. Cell. Biol.*, 23(5):1623–1632, Mar 2003.

[50] T. Manke, R. Bringas, and M. Vingron. Correlating protein-DNA and protein-protein interaction networks. *J. Mol. Biol.*, 333(1):75–85, Oct 2003.

[51] J. Göke, M. Jung, S. Behrens, L. Chavez, S. O'Keeffe, B. Timmermann, H. Lehrach, J. Adjaye, and M. Vingron. Combinatorial binding in human and mouse embryonic stem cells identifies conserved enhancers active in early embryonic development. *PLoS Comput. Biol.*, 7(12):e1002304, Dec 2011.

[52] M. Hemberg and G. Kreiman. Conservation of transcription factor binding events predicts gene expression across species. *Nucleic Acids Res.*, 39(16):7092–7102, Sep 2011.

[53] Q. He, A. F. Bardet, B. Patton, J. Purvis, J. Johnston, A. Paulson, M. Gogol, A. Stark, and J. Zeitlinger. High conservation of transcription factor binding and evidence for combinatorial regulation across six Drosophila species. *Nat. Genet.*, 43(5):414–420, May 2011.

[54] R. De Smet and K. Marchal. Advantages and limitations of current network inference methods. *Nat. Rev. Microbiol.*, 8(10):717–729, Oct 2010.

[55] L. A. Cirillo, F. R. Lin, I. Cuesta, D. Friedman, M. Jarnik, and K. S. Zaret. Opening of compacted chromatin by early developmental transcription factors HNF3 (FoxA) and GATA-4. *Mol. Cell*, 9(2):279–289, Feb 2002.

[56] A. A. Serandour, S. Avner, F. Percevault, F. Demay, M. Bizot, C. Lucchetti-Miganeh, F. Barloy-Hubler, M. Brown, M. Lupien, R. Metivier, G. Salbert, and J. Eeckhoute. Epigenetic switch involved in activation of pioneer factor FOXA1-dependent enhancers. *Genome Res.*, 21(4):555–565, Apr 2011.

[57] A. Soufi, G. Donahue, and K. S. Zaret. Facilitators and impediments of the pluripotency reprogramming factors' initial engagement with the genome. *Cell*, 151(5):994–1004, Nov 2012.

[58] T. S. Furey. ChIP-seq and beyond: new and improved methodologies to detect and characterize protein-DNA interactions. *Nat. Rev. Genet.*, 13(12):840–852, Dec 2012.

[59] I. Simon, J. Barnett, N. Hannett, C. T. Harbison, N. J. Rinaldi, T. L. Volkert, J. J. Wyrick, J. Zeitlinger, D. K. Gifford, T. S. Jaakkola, and R. A. Young. Serial regulation of transcriptional regulators in the yeast cell cycle. *Cell*, 106(6):697–708, Sep 2001.

[60] T. Ravasi, H. Suzuki, C. V. Cannistraci, S. Katayama, V. B. Bajic, K. Tan, A. Akalin, S. Schmeier, M. Kanamori-Katayama, N. Bertin, P. Carninci, C. O. Daub, A. R. Forrest, J. Gough, S. Grimmond, J. H. Han, T. Hashimoto, W. Hide, O. Hofmann, A. Kamburov, M. Kaur, H. Kawaji, A. Kubosaki, T. Lassmann, E. van Nimwegen, C. R. MacPherson, C. Ogawa, A. Radovanovic, A. Schwartz, R. D. Teasdale, J. Tegner, B. Lenhard, S. A. Teichmann, T. Arakawa, N. Ninomiya, K. Murakami, M. Tagami, S. Fukuda, K. Imamura, C. Kai, R. Ishihara, Y. Kitazume, J. Kawai, D. A. Hume, T. Ideker, and Y. Hayashizaki. An atlas of combinatorial transcriptional regulation in mouse and man. *Cell*, 140(5):744–752, Mar 2010.

[61] Satu Elisa Schaeffer. Survey: Graph clustering. *Comput. Sci. Rev.*, 1(1):27–64, August 2007.

[62] Vladimir Estivill-Castro. Why so many clustering algorithms: a position paper. *SIGKDD Explor. Newsl.*, 4(1):65–75, June 2002.

[63] F. Radicchi, C. Castellano, F. Cecconi, V. Loreto, and D. Parisi. Defining and identifying communities in networks. *Proc. Natl. Acad. Sci. U.S.A.*, 101(9):2658–2663, Mar 2004.

[64] D. J. Watts and S. H. Strogatz. Collective dynamics of 'small-world' networks. *Nature*, 393(6684):440–442, Jun 1998.

[65] P. Uetz, L. Giot, G. Cagney, T. A. Mansfield, R. S. Judson, J. R. Knight, D. Lockshon, V. Narayan, M. Srinivasan, P. Pochart, A. Qureshi-Emili, Y. Li, B. Godwin, D. Conover, T. Kalbfleisch, G. Vijayadamodar, M. Yang, M. Johnston, S. Fields, and J. M. Rothberg. A comprehensive analysis of protein-protein interactions in Saccharomyces cerevisiae. *Nature*, 403(6770):623–627, Feb 2000.

[66] T. Ito, T. Chiba, R. Ozawa, M. Yoshida, M. Hattori, and Y. Sakaki. A comprehensive two-hybrid analysis to explore the yeast protein interactome. *Proc. Natl. Acad. Sci. U.S.A.*, 98(8):4569–4574, Apr 2001.

[67] G. Rigaut, A. Shevchenko, B. Rutz, M. Wilm, M. Mann, and B. Seraphin. A generic protein purification method for protein complex characterization and proteome exploration. *Nat. Biotechnol.*, 17(10):1030–1032, Oct 1999.

[68] O. Puig, F. Caspary, G. Rigaut, B. Rutz, E. Bouveret, E. Bragado-Nilsson, M. Wilm, and B. Seraphin. The tandem affinity purification (TAP) method: a general procedure of protein complex purification. *Methods*, 24(3):218–229, Jul 2001.

[69] P. Aloy and R. B. Russell. Structural systems biology: modelling protein interactions. *Nat. Rev. Mol. Cell Biol.*, 7(3):188–197, Mar 2006.

[70] C. von Mering, R. Krause, B. Snel, M. Cornell, S. G. Oliver, S. Fields, and P. Bork. Comparative assessment of large-scale data sets of protein-protein interactions. *Nature*, 417(6887):399–403, May 2002.

[71] Q. C. Zhang, D. Petrey, J. I. Garzon, L. Deng, and B. Honig. PrePPI: a structure-informed database of protein-protein interactions. *Nucleic Acids Res.*, 41(Database issue):D828–833, Jan 2013.

[72] S. Pu, J. Wong, B. Turner, E. Cho, and S. J. Wodak. Up-to-date catalogues of yeast protein complexes. *Nucleic Acids Res.*, 37(3):825–831, Feb 2009.

[73] R. Saito, H. Suzuki, and Y. Hayashizaki. Interaction generality, a measurement to assess the reliability of a protein-protein interaction. *Nucleic Acids Res.*, 30(5):1163–1168, Mar 2002.

[74] C. M. Deane, L. Salwinski, I. Xenarios, and D. Eisenberg. Protein interactions: two methods for assessment of the reliability of high throughput observations. *Mol. Cell Proteomics*, 1(5):349–356, May 2002.

[75] E. Sprinzak, S. Sattath, and H. Margalit. How reliable are experimental protein-protein interaction data? *J. Mol. Biol.*, 327(5):919–923, Apr 2003.

[76] R. Jansen, H. Yu, D. Greenbaum, Y. Kluger, N. J. Krogan, S. Chung, A. Emili, M. Snyder, J. F. Greenblatt, and M. Gerstein. A Bayesian networks approach for predicting protein-protein interactions from genomic data. *Science*, 302(5644):449–453, Oct 2003.

[77] Q. C. Zhang, D. Petrey, L. Deng, L. Qiang, Y. Shi, C. A. Thu, B. Bisikirska, C. Lefebvre, D. Accili, T. Hunter, T. Maniatis, A. Califano, and B. Honig. Structure-based prediction of protein-protein interactions on a genome-wide scale. *Nature*, 490(7421):556–560, Oct 2012.

[78] L. H. Hartwell, J. J. Hopfield, S. Leibler, and A. W. Murray. From molecular to modular cell biology. *Nature*, 402(6761 Suppl):47–52, Dec 1999.

[79] N. Przulj, D. A. Wigle, and I. Jurisica. Functional topology in a network of protein interactions. *Bioinformatics*, 20(3):340–348, Feb 2004.

[80] G. D. Bader and C. W. Hogue. An automated method for finding molecular complexes in large protein interaction networks. *BMC Bioinformatics*, 4:2, Jan 2003.

[81] T. Nepusz, H. Yu, and A. Paccanaro. Detecting overlapping protein complexes in protein-protein interaction networks. *Nat. Methods*, 9(5):471–472, May 2012.

[82] D. Skowyra, K. L. Craig, M. Tyers, S. J. Elledge, and J. W. Harper. F-box proteins are receptors that recruit phosphorylated substrates to the SCF ubiquitin-ligase complex. *Cell*, 91(2):209–219, Oct 1997.

[83] S. H. Jung, B. Hyun, W. H. Jang, H. Y. Hur, and D. S. Han. Protein complex prediction based on simultaneous protein interaction network. *Bioinformatics*, 26(3):385–391, Feb 2010.

[84] J. D. Han, N. Bertin, T. Hao, D. S. Goldberg, G. F. Berriz, L. V. Zhang, D. Dupuy, A. J. Walhout, M. E. Cusick, F. P. Roth, and M. Vidal. Evidence for dynamically organized modularity in the yeast protein-protein interaction network. *Nature*, 430(6995):88–93, Jul 2004.

[85] P. M. Kim, L. J. Lu, Y. Xia, and M. B. Gerstein. Relating three-dimensional structures to protein networks provides evolutionary insights. *Science*, 314(5807):1938–1941, Dec 2006.

[86] O. Keskin and R. Nussinov. Similar binding sites and different partners: implications to shared proteins in cellular pathways. *Structure*, 15(3):341–354, Mar 2007.

[87] X. L. Li, S. H. Tan, C. S. Foo, and S. K. Ng. Interaction graph mining for protein complexes using local clique merging. *Genome Inform*, 16(2):260–269, 2005.

[88] A. D. King, N. Przulj, and I. Jurisica. Protein complex prediction via cost-based clustering. *Bioinformatics*, 20(17):3013–3020, Nov 2004.

[89] A. J. Enright, S. Van Dongen, and C. A. Ouzounis. An efficient algorithm for large-scale detection of protein families. *Nucleic Acids Res.*, 30(7):1575–1584, Apr 2002.

[90] K. Macropol, T. Can, and A. K. Singh. RRW: repeated random walks on genome-scale protein networks for local cluster discovery. *BMC Bioinformatics*, 10:283, 2009.

[91] M. Habibi, C. Eslahchi, and L. Wong. Protein complex prediction based on k-connected subgraphs in protein interaction network. *BMC Syst Biol*, 4:129, 2010.

[92] S. Brohee and J. van Helden. Evaluation of clustering algorithms for protein-protein interaction networks. *BMC Bioinformatics*, 7:488, 2006.

[93] Y. Ozawa, R. Saito, S. Fujimori, H. Kashima, M. Ishizaka, H. Yana-gawa, E. Miyamoto-Sato, and M. Tomita. Protein complex prediction via verifying and reconstructing the topology of domain-domain in-teractions. *BMC Bioinformatics*, 11:350, 2010.

[94] W. Ma, C. McAnulla, and L. Wang. Protein complex prediction based on maximum matching with domain-domain interaction. *Biochim. Biophys. Acta*, 1824(12):1418–1424, Dec 2012.

[95] X. Li, M. Wu, C. K. Kwoh, and S. K. Ng. Computational approaches for detecting protein complexes from protein interaction networks: a survey. *BMC Genomics*, 11 Suppl 1:S3, 2010.

[96] C. J. Van Rijsbergen. *Information Retrieval.* Butterworth-Heinemann, Newton, MA, USA, 2nd edition, 1979.

[97] Thomas H. Cormen, Charles E. Leiserson, Ronald L. Rivest, and Clif-ford Stein. *Introduction to Algorithms.* The MIT Press, 3rd edition, 2009.

[98] C. C. Friedel, J. Krumsiek, and R. Zimmer. Bootstrapping the inter-actome: unsupervised identification of protein complexes in yeast. *J. Comput. Biol.*, 16(8):971–987, Aug 2009.

[99] M. Ashburner, C. A. Ball, J. A. Blake, D. Botstein, H. Butler, J. M. Cherry, A. P. Davis, K. Dolinski, S. S. Dwight, J. T. Eppig, M. A. Harris, D. P. Hill, L. Issel-Tarver, A. Kasarskis, S. Lewis, J. C. Matese, J. E. Richardson, M. Ringwald, G. M. Rubin, and G. Sher-lock. Gene ontology: tool for the unification of biology. The Gene Ontology Consortium. *Nat. Genet.*, 25(1):25–29, May 2000.

[100] B. Zhang, B. H. Park, T. Karpinets, and N. F. Samatova. From pull-down data to protein interaction networks and complexes with biological relevance. *Bioinformatics*, 24(7):979–986, Apr 2008.

[101] D. Bu, Y. Zhao, L. Cai, H. Xue, X. Zhu, H. Lu, J. Zhang, S. Sun, L. Ling, N. Zhang, G. Li, and R. Chen. Topological structure analysis of the protein-protein interaction network in budding yeast. *Nucleic Acids Res.*, 31(9):2443–2450, May 2003.

[102] T. Hastie, R. Tibshirani, and J. Friedman. *The elements of statistical learning: data mining, inference and prediction.* Springer, 2 edition, 2009.

[103] R. Jansen, D. Greenbaum, and M. Gerstein. Relating whole-genome expression data with protein-protein interactions. *Genome Res.*, 12(1):37–46, Jan 2002.

[104] V. Spirin and L. A. Mirny. Protein complexes and functional modules in molecular networks. *Proc. Natl. Acad. Sci. U.S.A.*, 100(21):12123–12128, Oct 2003.

[105] S. Gong, G. Yoon, I. Jang, D. Bolser, P. Dafas, M. Schroeder, H. Choi, Y. Cho, K. Han, S. Lee, H. Choi, M. Lappe, L. Holm, S. Kim, D. Oh, and J. Bhak. PSIbase: a database of Protein Structural Interactome map (PSIMAP). *Bioinformatics*, 21(10):2541–2543, May 2005.

[106] E. Sprinzak, Y. Altuvia, and H. Margalit. Characterization and prediction of protein-protein interactions within and between complexes. *Proc. Natl. Acad. Sci. U.S.A.*, 103(40):14718–14723, Oct 2006.

[107] K. S. Guimaraes, R. Jothi, E. Zotenko, and T. M. Przytycka. Predicting domain-domain interactions using a parsimony approach. *Genome Biol.*, 7(11):R104, 2006.

[108] L. Zhao, S. C. Hoi, L. Wong, T. Hamp, and J. Li. Structural and functional analysis of multi-interface domains. *PLoS ONE*, 7(12):e50821, 2012.

[109] E. L. Sonnhammer, S. R. Eddy, and R. Durbin. Pfam: a comprehensive database of protein domain families based on seed alignments. *Proteins*, 28(3):405–420, Jul 1997.

[110] R. Apweiler, T. K. Attwood, A. Bairoch, A. Bateman, E. Birney, M. Biswas, P. Bucher, L. Cerutti, F. Corpet, M. D. Croning, R. Durbin, L. Falquet, W. Fleischmann, J. Gouzy, H. Hermjakob, N. Hulo, I. Jonassen, D. Kahn, A. Kanapin, Y. Karavidopoulou, R. Lopez, B. Marx, N. J. Mulder, T. M. Oinn, M. Pagni, F. Servant, C. J. Sigrist, and E. M. Zdobnov. InterPro–an integrated documentation resource for protein families, domains and functional sites. *Bioinformatics*, 16(12):1145–1150, Dec 2000.

[111] Y. Kim, B. Min, and G. S. Yi. IDDI: integrated domain-domain interaction and protein interaction analysis system. *Proteome Sci*, 10 Suppl 1:S9, 2012.

[112] S. Yellaboina, A. Tasneem, D. V. Zaykin, B. Raghavachari, and R. Jothi. DOMINE: a comprehensive collection of known and predicted domain-domain interactions. *Nucleic Acids Res.*, 39(Database issue):D730–735, Jan 2011.

[113] A. Schrijver. *Theory of Linear and Integer Programming.* Wiley Series in Discrete Mathematics & Optimization. John Wiley & Sons, 1998.

[114] J. Edmonds. Paths, trees, and flowers. *Canadian Journal of Mathematics*, 17:449–467, Feb 1965.

[115] P. Bjorkholm and E. L. Sonnhammer. Comparative analysis and unification of domain-domain interaction networks. *Bioinformatics*, 25(22):3020–3025, Nov 2009.

[116] Y. Pilpel, P. Sudarsanam, and G. M. Church. Identifying regulatory networks by combinatorial analysis of promoter elements. *Nat. Genet.*, 29(2):153–159, Oct 2001.

[117] Y. H. Chang, Y. C. Wang, and B. S. Chen. Identification of transcription factor cooperativity via stochastic system model. *Bioinformatics*, 22(18):2276–2282, Sep 2006.

[118] N. Banerjee and M. Q. Zhang. Identifying cooperativity among transcription factors controlling the cell cycle in yeast. *Nucleic Acids Res.*, 31(23):7024–7031, Dec 2003.

[119] N. Nagamine, Y. Kawada, and Y. Sakakibara. Identifying cooperative transcriptional regulations using protein-protein interactions. *Nucleic Acids Res.*, 33(15):4828–4837, 2005.

[120] H. K. Tsai, H. H. Lu, and W. H. Li. Statistical methods for identifying yeast cell cycle transcription factors. *Proc. Natl. Acad. Sci. U.S.A.*, 102(38):13532–13537, Sep 2005.

[121] X. Yu, J. Lin, T. Masuda, N. Esumi, D. J. Zack, and J. Qian. Genome-wide prediction and characterization of interactions between transcription factors in Saccharomyces cerevisiae. *Nucleic Acids Res.*, 34(3):917–927, 2006.

[122] G. Wang and W. Zhang. A steganalysis-based approach to comprehensive identification and characterization of functional regulatory elements. *Genome Biol.*, 7(6):R49, 2006.

[123] M. Lapidot and Y. Pilpel. Comprehensive quantitative analyses of the effects of promoter sequence elements on mRNA transcription. *Nucleic Acids Res.*, 31(13):3824–3828, Jul 2003.

[124] L. Song, P. Langfelder, and S. Horvath. Comparison of co-expression measures: mutual information, correlation, and model based indices. *BMC Bioinformatics*, 13:328, 2012.

[125] J. Bang-Jensen, G. Gutin, and A. Yeo. When the greedy algorithm fails. *Discrete Optimization*, 1(2):121 – 127, 2004.

[126] R. Apweiler, A. Bairoch, C. H. Wu, W. C. Barker, B. Boeckmann, S. Ferro, E. Gasteiger, H. Huang, R. Lopez, M. Magrane, M. J. Martin, D. A. Natale, C. O'Donovan, N. Redaschi, and L. S. Yeh. UniProt: the Universal Protein knowledgebase. *Nucleic Acids Res.*, 32(Database issue):D115–119, Jan 2004.

[127] The UniProt Consortium and Apweiler R. Update on activities at the Universal Protein Resource (UniProt) in 2013. *Nucleic Acids Res.*, 41(Database issue):D43–47, Jan 2013.

[128] S. Hunter, P. Jones, A. Mitchell, R. Apweiler, T. K. Attwood, A. Bateman, T. Bernard, D. Binns, P. Bork, S. Burge, E. de Castro, P. Coggill, M. Corbett, U. Das, L. Daugherty, L. Duquenne, R. D. Finn, M. Fraser, J. Gough, D. Haft, N. Hulo, D. Kahn, E. Kelly, I. Letunic, D. Lonsdale, R. Lopez, M. Madera, J. Maslen, C. McAnulla, J. McDowall, C. McMenamin, H. Mi, P. Mutowo-Muellenet, N. Mulder, D. Natale, C. Orengo, S. Pesseat, M. Punta, A. F. Quinn, C. Rivoire, A. Sangrador-Vegas, J. D. Selengut, C. J. Sigrist, M. Scheremetjew, J. Tate, M. Thimmajanarthanan, P. D. Thomas, C. H. Wu, C. Yeats, and S. Y. Yong. InterPro in 2011: new developments in the family and domain prediction database. *Nucleic Acids Res.*, 40(Database issue):D306–312, Jan 2012.

[129] Guido van Rossum. Python reference manual. Technical report, Amsterdam, The Netherlands, 1995.

[130] H. W. Mewes, D. Frishman, K. F. Mayer, M. Munsterkotter, O. Noubibou, P. Pagel, T. Rattei, M. Oesterheld, A. Ruepp, and V. Stumpflen. MIPS: analysis and annotation of proteins from whole genomes in 2005. *Nucleic Acids Res.*, 34(Database issue):D169–172, Jan 2006.

[131] I. Xenarios, D. W. Rice, L. Salwinski, M. K. Baron, E. M. Marcotte, and D. Eisenberg. DIP: the database of interacting proteins. *Nucleic Acids Res.*, 28(1):289–291, Jan 2000.

[132] S. Kerrien, Y. Alam-Faruque, B. Aranda, I. Bancarz, A. Bridge, C. Derow, E. Dimmer, M. Feuermann, A. Friedrichsen, R. Huntley, C. Kohler, J. Khadake, C. Leroy, A. Liban, C. Lieftink, L. Montecchi-Palazzi, S. Orchard, J. Risse, K. Robbe, B. Roechert, D. Thorneycroft, Y. Zhang, R. Apweiler, and H. Hermjakob. IntAct–open source resource for molecular interaction data. *Nucleic Acids Res.*, 35(Database issue):D561–565, Jan 2007.

[133] A. Zanzoni, L. Montecchi-Palazzi, M. Quondam, G. Ausiello, M. Helmer-Citterich, and G. Cesareni. MINT: a Molecular INTeraction database. *FEBS Lett.*, 513(1):135–140, Feb 2002.

[134] T. S. Keshava Prasad, R. Goel, K. Kandasamy, S. Keerthikumar, S. Kumar, S. Mathivanan, D. Telikicherla, R. Raju, B. Shafreen, A. Venugopal, L. Balakrishnan, A. Marimuthu, S. Banerjee, D. S. Somanathan, A. Sebastian, S. Rani, S. Ray, C. J. Harrys Kishore, S. Kanth, M. Ahmed, M. K. Kashyap, R. Mohmood, Y. L. Ramachandra, V. Krishna, B. A. Rahiman, S. Mohan, P. Ranganathan, S. Ramabadran, R. Chaerkady, and A. Pandey. Human Protein Reference Database–2009 update. *Nucleic Acids Res.*, 37(Database issue):D767–772, Jan 2009.

[135] C. Stark, B. J. Breitkreutz, T. Reguly, L. Boucher, A. Breitkreutz, and M. Tyers. BioGRID: a general repository for interaction datasets. *Nucleic Acids Res.*, 34(Database issue):D535–539, Jan 2006.

[136] A. Kumar and M. Snyder. Protein complexes take the bait. *Nature*, 415(6868):123–124, Jan 2002.

[137] C. von Mering, R. Krause, B. Snel, M. Cornell, S. G. Oliver, S. Fields, and P. Bork. Comparative assessment of large-scale data sets of protein-protein interactions. *Nature*, 417(6887):399–403, May 2002.

[138] R. Saito, M. E. Smoot, K. Ono, J. Ruscheinski, P. L. Wang, S. Lotia, A. R. Pico, G. D. Bader, and T. Ideker. A travel guide to Cytoscape plugins. *Nat. Methods*, 9(11):1069–1076, Nov 2012.

[139] R. Durbin, S. Eddy, A. Krogh, and G. Mitchison. *Biological sequence analysis*. Cambridge University Press, 14th edition, 2010.

[140] A. Krogh, M. Brown, I. S. Mian, K. Sjolander, and D. Haussler. Hidden Markov models in computational biology. Applications to protein modeling. *J. Mol. Biol.*, 235(5):1501–1531, Feb 1994.

[141] A. Bateman and D. H. Haft. HMM-based databases in InterPro. *Brief. Bioinformatics*, 3(3):236–245, Sep 2002.

[142] M. Punta, P. C. Coggill, R. Y. Eberhardt, J. Mistry, J. Tate, C. Boursnell, N. Pang, K. Forslund, G. Ceric, J. Clements, A. Heger, L. Holm, E. L. Sonnhammer, S. R. Eddy, A. Bateman, and R. D. Finn. The Pfam protein families database. *Nucleic Acids Res.*, 40(Database issue):290–301, Jan 2012.

[143] E. L. Sonnhammer and D. Kahn. Modular arrangement of proteins as inferred from analysis of homology. *Protein Sci.*, 3(3):482–492, Mar 1994.

[144] K. D. Pruitt, T. Tatusova, G. R. Brown, and D. R. Maglott. NCBI Reference Sequences (RefSeq): current status, new features and genome annotation policy. *Nucleic Acids Res.*, 40(Database issue):D130–135, Jan 2012.

[145] T. K. Attwood, A. Coletta, G. Muirhead, A. Pavlopoulou, P. B. Philippou, I. Popov, C. Roma-Mateo, A. Theodosiou, and A. L. Mitchell. The PRINTS database: a fine-grained protein sequence annotation and analysis resource–its status in 2012. *Database (Oxford)*, 2012:bas019, 2012.

[146] C. J. Sigrist, E. de Castro, L. Cerutti, B. A. Cuche, N. Hulo, A. Bridge, L. Bougueleret, and I. Xenarios. New and continuing developments at PROSITE. *Nucleic Acids Res.*, 41(Database issue):D344–347, Jan 2013.

[147] I. Letunic, T. Doerks, and P. Bork. SMART 7: recent updates to the protein domain annotation resource. *Nucleic Acids Res.*, 40(Database issue):D302–305, Jan 2012.

[148] F. Servant, C. Bru, S. Carrere, E. Courcelle, J. Gouzy, D. Peyruc, and D. Kahn. ProDom: automated clustering of homologous domains. *Brief. Bioinformatics*, 3(3):246–251, Sep 2002.

[149] S. F. Altschul, T. L. Madden, A. A. Schaffer, J. Zhang, Z. Zhang, W. Miller, and D. J. Lipman. Gapped BLAST and PSI-BLAST: a new generation of protein database search programs. *Nucleic Acids Res.*, 25(17):3389–3402, Sep 1997.

[150] A. G. Murzin, S. E. Brenner, T. Hubbard, and C. Chothia. SCOP: a structural classification of proteins database for the investigation of sequences and structures. *J. Mol. Biol.*, 247(4):536–540, Apr 1995.

[151] A. N. Nikolskaya, C. N. Arighi, H. Huang, W. C. Barker, and C. H. Wu. PIRSF family classification system for protein functional and evolutionary analysis. *Evol. Bioinform. Online*, 2:197–209, 2006.

[152] D. Wilson, R. Pethica, Y. Zhou, C. Talbot, C. Vogel, M. Madera, C. Chothia, and J. Gough. SUPERFAMILY–sophisticated comparative genomics, data mining, visualization and phylogeny. *Nucleic Acids Res.*, 37(Database issue):D380–386, Jan 2009.

[153] H. Mi, A. Muruganujan, and P. D. Thomas. PANTHER in 2013: modeling the evolution of gene function, and other gene attributes, in the context of phylogenetic trees. *Nucleic Acids Res.*, 41(Database issue):D377–386, Jan 2013.

[154] J. Lees, C. Yeats, J. Perkins, I. Sillitoe, R. Rentzsch, B. H. Dessailly, and C. Orengo. Gene3D: a domain-based resource for comparative genomics, functional annotation and protein network analysis. *Nucleic Acids Res.*, 40(Database issue):D465–471, Jan 2012.

[155] C. A. Orengo, A. D. Michie, S. Jones, D. T. Jones, M. B. Swindells, and J. M. Thornton. CATH–a hierarchic classification of protein domain structures. *Structure*, 5(8):1093–1108, Aug 1997.

[156] D. H. Haft, J. D. Selengut, R. A. Richter, D. Harkins, M. K. Basu, and E. Beck. TIGRFAMs and Genome Properties in 2013. *Nucleic Acids Res.*, 41(Database issue):D387–395, Jan 2013.

[157] I. Pedruzzi, C. Rivoire, A. H. Auchincloss, E. Coudert, G. Keller, E. de Castro, D. Baratin, B. A. Cuche, L. Bougueleret, S. Poux,

N. Redaschi, I. Xenarios, and A. Bridge. HAMAP in 2013, new developments in the protein family classification and annotation system. *Nucleic Acids Res.*, 41(Database issue):D584–589, Jan 2013.

[158] S. Hunter, R. Apweiler, T. K. Attwood, A. Bairoch, A. Bateman, D. Binns, P. Bork, U. Das, L. Daugherty, L. Duquenne, R. D. Finn, J. Gough, D. Haft, N. Hulo, D. Kahn, E. Kelly, A. Laugraud, I. Letunic, D. Lonsdale, R. Lopez, M. Madera, J. Maslen, C. McAnulla, J. McDowall, J. Mistry, A. Mitchell, N. Mulder, D. Natale, C. Orengo, A. F. Quinn, J. D. Selengut, C. J. Sigrist, M. Thimma, P. D. Thomas, F. Valentin, D. Wilson, C. H. Wu, and C. Yeats. InterPro: the integrative protein signature database. *Nucleic Acids Res.*, 37(Database issue):D211–215, Jan 2009.

[159] E. M. Zdobnov and R. Apweiler. InterProScan–an integration platform for the signature-recognition methods in InterPro. *Bioinformatics*, 17(9):847–848, Sep 2001.

[160] R. Lopez, K. Duggan, N. Harte, and A. Kibria. Public services from the European Bioinformatics Institute. *Brief. Bioinformatics*, 4(4):332–340, Dec 2003.

[161] D. T. Chang, C. Y. Huang, C. Y. Wu, and W. S. Wu. YPA: an integrated repository of promoter features in Saccharomyces cerevisiae. *Nucleic Acids Res.*, 39(Database issue):D647–652, Jan 2011.

[162] K. D. MacIsaac, T. Wang, D. B. Gordon, D. K. Gifford, G. D. Stormo, and E. Fraenkel. An improved map of conserved regulatory sites for Saccharomyces cerevisiae. *BMC Bioinformatics*, 7:113, 2006.

[163] H. K. Tsai, M. Y. Chou, C. H. Shih, G. T. Huang, T. H. Chang, and W. H. Li. MYBS: a comprehensive web server for mining transcription factor binding sites in yeast. *Nucleic Acids Res.*, 35(Web Server issue):W221–226, Jul 2007.

[164] M. Pachkov, P. J. Balwierz, P. Arnold, E. Ozonov, and E. van Nimwegen. SwissRegulon, a database of genome-wide annotations of regulatory sites: recent updates. *Nucleic Acids Res.*, 41(Database issue):D214–220, Jan 2013.

[165] D. Abdulrehman, P. T. Monteiro, M. C. Teixeira, N. P. Mira, A. B. Lourenco, S. C. dos Santos, T. R. Cabrito, A. P. Francisco, S. C.

Madeira, R. S. Aires, A. L. Oliveira, I. Sa-Correia, and A. T. Freitas. YEASTRACT: providing a programmatic access to curated transcriptional regulatory associations in Saccharomyces cerevisiae through a web services interface. *Nucleic Acids Res.*, 39(Database issue):D136–140, Jan 2011.

[166] J. Zhu and M. Q. Zhang. SCPD: a promoter database of the yeast Saccharomyces cerevisiae. *Bioinformatics*, 15(7-8):607–611, 1999.

[167] C. Zhu, K. J. Byers, R. P. McCord, Z. Shi, M. F. Berger, D. E. Newburger, K. Saulrieta, Z. Smith, M. V. Shah, M. Radhakrishnan, A. A. Philippakis, Y. Hu, F. De Masi, M. Pacek, A. Rolfs, T. Murthy, J. Labaer, and M. L. Bulyk. High-resolution DNA-binding specificity analysis of yeast transcription factors. *Genome Res.*, 19(4):556–566, Apr 2009.

[168] M. A. Garcia-Gimeno and K. Struhl. Aca1 and Aca2, ATF/CREB activators in Saccharomyces cerevisiae, are important for carbon source utilization but not the response to stress. *Mol. Cell. Biol.*, 20(12):4340–4349, Jun 2000.

[169] J. O. Nehlin, M. Carlberg, and H. Ronne. Yeast SKO1 gene encodes a bZIP protein that binds to the CRE motif and acts as a repressor of transcription. *Nucleic Acids Res.*, 20(20):5271–5278, Oct 1992.

[170] L. Dirick, T. Moll, H. Auer, and K. Nasmyth. A central role for SWI6 in modulating cell cycle Start-specific transcription in yeast. *Nature*, 357(6378):508–513, Jun 1992.

[171] B. J. Andrews and L. A. Moore. Interaction of the yeast Swi4 and Swi6 cell cycle regulatory proteins in vitro. *Proc. Natl. Acad. Sci. U.S.A.*, 89(24):11852–11856, Dec 1992.

[172] O. Vincent and M. Carlson. Sip4, a Snf1 kinase-dependent transcriptional activator, binds to the carbon source-responsive element of gluconeogenic genes. *EMBO J.*, 17(23):7002–7008, Dec 1998.

[173] S. L. Chin, I. M. Marcus, R. R. Klevecz, and C. M. Li. Dynamics of oscillatory phenotypes in Saccharomyces cerevisiae reveal a network of genome-wide transcriptional oscillators. *FEBS J.*, 279(6):1119–1130, Mar 2012.

[174] R. Edgar, M. Domrachev, and A. E. Lash. Gene Expression Omnibus: NCBI gene expression and hybridization array data repository. *Nucleic Acids Res.*, 30(1):207–210, Jan 2002.

[175] R Core Team. *R: A Language and Environment for Statistical Computing*. R Foundation for Statistical Computing, Vienna, Austria, 2013.

[176] Robert C Gentleman, Vincent J. Carey, Douglas M. Bates, and others. Bioconductor: Open software development for computational biology and bioinformatics. *Genome Biology*, 5:R80, 2004.

[177] Z. J. Wu, R. A. Irizarry, R. Gentleman, F. Martinez-Murillo, and F. Spencer. A model-based background adjustment for oligonucleotide expression arrays. *J. Am. Stat. Assoc.*, 99(468):909–917, Dec 2004.

[178] E. Hubbell, W. M. Liu, and R. Mei. Robust estimators for expression analysis. *Bioinformatics*, 18(12):1585–1592, Dec 2002.

[179] W. K. Lim, K. Wang, C. Lefebvre, and A. Califano. Comparative analysis of microarray normalization procedures: effects on reverse engineering gene networks. *Bioinformatics*, 23(13):i282–288, Jul 2007.

[180] Laurent Gautier, Leslie Cope, Benjamin M. Bolstad, and Rafael A. Irizarry. affy—analysis of affymetrix genechip data at the probe level. *Bioinformatics*, 20(3):307–315, 2004.

[181] Marc Carlson. *yeast2.db: Affymetrix Yeast Genome 2.0 Array annotation data (chip yeast2)*. R package version 2.9.0.

[182] L. J. Heyer, S. Kruglyak, and S. Yooseph. Exploring expression data: identification and analysis of coexpressed genes. *Genome Res.*, 9(11):1106–1115, Nov 1999.

[183] J. M. Cherry, E. L. Hong, C. Amundsen, R. Balakrishnan, G. Binkley, E. T. Chan, K. R. Christie, M. C. Costanzo, S. S. Dwight, S. R. Engel, D. G. Fisk, J. E. Hirschman, B. C. Hitz, K. Karra, C. J. Krieger, S. R. Miyasato, R. S. Nash, J. Park, M. S. Skrzypek, M. Simison, S. Weng, and E. D. Wong. Saccharomyces Genome Database: the genomics resource of budding yeast. *Nucleic Acids Res.*, 40(Database issue):D700–705, Jan 2012.

[184] W. K. Huh, J. V. Falvo, L. C. Gerke, A. S. Carroll, R. W. How-
son, J. S. Weissman, and E. K. O'Shea. Global analysis of protein
localization in budding yeast. *Nature*, 425(6959):686–691, Oct 2003.

[185] Inside the Python GIL. In D. Beazley, editor, *Python Concurrency
Workshop*, Chicago, 2009.

[186] Aric A. Hagberg, Daniel A. Schult, and Pieter J. Swart. Exploring
network structure, dynamics, and function using NetworkX. In *Pro-
ceedings of the 7th Python in Science Conference (SciPy2008)*, pages
11–15, Pasadena, CA USA, Aug 2008.

[187] Eric Jones, Travis Oliphant, Pearu Peterson, et al. SciPy: Open
source scientific tools for Python, 2001–.

[188] H. Tang, B. Petersen, A. Naldi, and P. Flick. goatools: Python scripts
to find enrichment of GO terms.

[189] J. D. Hunter. Matplotlib: A 2d graphics environment. *Computing In
Science & Engineering*, 9(3):90–95, 2007.

[190] S. R. Engel, F. S. Dietrich, D. G. Fisk, G. Binkley, R. Balakrishnan,
M. C. Costanzo, S. S. Dwight, B. C. Hitz, K. Karra, R. S. Nash,
S. Weng, E. D. Wong, P. Lloyd, M. S. Skrzypek, S. R. Miyasato,
M. Simison, and J. M. Cherry. The Reference Genome Sequence of
Saccharomyces cerevisiae: Then and Now. *G3 (Bethesda)*, Dec 2013.

[191] X. Xin, C. Lan, H. C. Lee, and L. Zhang. Regulation of the HAP1
gene involves positive actions of histone deacetylases. *Biochem. Biophys. Res. Commun.*, 362(1):120–125, Oct 2007.

[192] D. Binns, E. Dimmer, R. Huntley, D. Barrell, C. O'Donovan, and
R. Apweiler. QuickGO: a web-based tool for Gene Ontology search-
ing. *Bioinformatics*, 25(22):3045–3046, Nov 2009.

[193] Z. Zhang and J. C. Reese. Molecular genetic analysis of the yeast
repressor Rfx1/Crt1 reveals a novel two-step regulatory mechanism.
Mol. Cell. Biol., 25(17):7399–7411, Sep 2005.

[194] L. Bai, A. Ondracka, and F. R. Cross. Multiple sequence-specific
factors generate the nucleosome-depleted region on CLN2 promoter.
Mol. Cell, 42(4):465–476, May 2011.

[195] D. E. Martin, A. Soulard, and M. N. Hall. TOR regulates ribosomal protein gene expression via PKA and the Forkhead transcription factor FHL1. *Cell*, 119(7):969–979, Dec 2004.

[196] K. A. Olson, C. Nelson, G. Tai, W. Hung, C. Yong, C. Astell, and I. Sadowski. Two regulators of Ste12p inhibit pheromone-responsive transcription by separate mechanisms. *Mol. Cell. Biol.*, 20(12):4199–4209, Jun 2000.

[197] Y. Zhao, K. B. McIntosh, D. Rudra, S. Schawalder, D. Shore, and J. R. Warner. Fine-structure analysis of ribosomal protein gene transcription. *Mol. Cell. Biol.*, 26(13):4853–4862, Jul 2006.

[198] V. V. Svetlov and T. G. Cooper. The Saccharomyces cerevisiae GATA factors Dal80p and Deh1p can form homo- and heterodimeric complexes. *J. Bacteriol.*, 180(21):5682–5688, Nov 1998.

[199] T. S. Cunningham, R. A. Dorrington, and T. G. Cooper. The UGA4 UASNTR site required for GLN3-dependent transcriptional activation also mediates DAL80-responsive regulation and DAL80 protein binding in Saccharomyces cerevisiae. *J. Bacteriol.*, 176(15):4718–4725, Aug 1994.

[200] C. Devlin, K. Tice-Baldwin, D. Shore, and K. T. Arndt. RAP1 is required for BAS1/BAS2- and GCN4-dependent transcription of the yeast HIS4 gene. *Mol. Cell. Biol.*, 11(7):3642–3651, Jul 1991.

[201] J. Orzechowski Westholm, S. Tronnersjo, N. Nordberg, I. Olsson, J. Komorowski, and H. Ronne. Gis1 and Rph1 regulate glycerol and acetate metabolism in glucose depleted yeast cells. *PLoS ONE*, 7(2):e31577, 2012.

[202] E. M. Kim, Y. K. Jang, and S. D. Park. Phosphorylation of Rph1, a damage-responsive repressor of PHR1 in Saccharomyces cerevisiae, is dependent upon Rad53 kinase. *Nucleic Acids Res.*, 30(3):643–648, Feb 2002.

[203] V. R. Iyer, C. E. Horak, C. S. Scafe, D. Botstein, M. Snyder, and P. O. Brown. Genomic binding sites of the yeast cell-cycle transcription factors SBF and MBF. *Nature*, 409(6819):533–538, Jan 2001.

[204] M. Costanzo, O. Schub, and B. Andrews. G1 transcription factors are differentially regulated in Saccharomyces cerevisiae by the Swi6-binding protein Stb1. *Mol. Cell. Biol.*, 23(14):5064–5077, Jul 2003.

[205] J. M. Bean, E. D. Siggia, and F. R. Cross. High functional overlap between MluI cell-cycle box binding factor and Swi4/6 cell-cycle box binding factor in the G1/S transcriptional program in Saccharomyces cerevisiae. *Genetics*, 171(1):49–61, Sep 2005.

[206] D. A. Orlando, C. Y. Lin, A. Bernard, J. Y. Wang, J. E. Socolar, E. S. Iversen, A. J. Hartemink, and S. B. Haase. Global control of cell-cycle transcription by coupled CDK and network oscillators. *Nature*, 453(7197):944–947, Jun 2008.

[207] D. S. McNabb and I. Pinto. Assembly of the Hap2p/Hap3p/Hap4p/Hap5p-DNA complex in Saccharomyces cerevisiae. *Eukaryotic Cell*, 4(11):1829–1839, Nov 2005.

[208] A. Shevchenko, A. Roguev, D. Schaft, L. Buchanan, B. Habermann, C. Sakalar, H. Thomas, N. J. Krogan, A. Shevchenko, and A. F. Stewart. Chromatin Central: towards the comparative proteome by accurate mapping of the yeast proteomic environment. *Genome Biol.*, 9(11):R167, 2008.